Principles of Fluid Mechanics

Jürgen Zierep · Karl Bühler

Principles of Fluid Mechanics

Fundamentals, Statics and Dynamics of Fluids

 Springer

Jürgen Zierep
Karlsruhe, Germany

Karl Bühler
Hochschule Offenburg
Offenburg, Germany

This book is a translation of the original German edition "Grundzüge der Strömungslehre" by Zierep, Jürgen and Bühler, Karl, published by Springer Fachmedien Wiesbaden GmbH in 2018. The translation was done with the help of artificial intelligence (machine translation by the service DeepL.com). A subsequent human revision was done primarily in terms of content, so that the book will read stylistically differently from a conventional translation. Springer Nature works continuously to further the development of tools for the production of books and on the related technologies to support the authors.

ISBN 978-3-658-34811-3   ISBN 978-3-658-34812-0   (eBook)
https://doi.org/10.1007/978-3-658-34812-0

This Springer imprint is published by the registered company Springer Fachmedien Wiesbaden GmbH part of Springer Nature.
The registered company address is: Abraham-Lincoln-Str. 46, 65189 Wiesbaden, Germany

Preface to the 11th Edition

The *Principles of Fluid Mechanics* (Grundzüge der Strömungslehre) has proved very popular with teachers and students over the past 40 years through its 10 editions.

We have been able to ascertain this time and again on the basis of lectures and discussions at home and abroad. The statements of the preface of the first edition—especially concerning the application of the important momentum theorem—are still valid.

"First, to gain knowledge on examples of fluid mechanics and then, second, to consider the control space and the flow data on the boundary" are important prerequisites for the successful application of the momentum theorem.

Over time, the desire arose, e.g., during lectures at the TH Budapest, for further current exercises in fluid mechanics. Preference is given to problems with the following alternative: inflow or outflow from containers, without or with friction, stationary or unsteady, incompressible or compressible. The implementation of the conservation laws for mass, momentum, and energy based on the initial and boundary conditions in concrete problems is still often difficult.

The energy theorem is central to many flow problems today. This is illustrated in the following examples. The energy theorem is derived and applied in many forms. Interesting and typical are the values in the Rayleigh–Stokes problem, where the time variation of kinetic energy, dissipation, and wall shear stress power occur simultaneously. In the potential flows of viscous media, the physical processes take place on the energetic side. The physical principles of these viscous potential flows are covered in the text and the exercises on vortex flow and source–sink flow in the cylinder gap. These statements play an increasingly important role today where the energy balance is concerned. Up to now, they have been little represented in the textbooks of fluid mechanics.

The book is intended for students of engineering and physics at universities and colleges to accompany and reinforce lectures on fluid mechanics and for self-study. This book is also useful for engineers working in practice as an introduction to and consolidation of fluid mechanics fundamentals.

We have dealt with some relevant problems of this kind in detail and added them at the end of the book. Again, it is the case that the reader must often reach for paper and pencil in order to be able to follow the solution path proposed by us.

We would like to thank the mechanical engineering editorial office of Springer Vieweg for their active support and great confidence in the publication of this book and the extremely gratifying cooperation.

Karlsruhe, Germany   Jürgen Zierep
November 2017   Karl Bühler

From the Preface of the 1st Edition

The present book *Principles of Fluid Mechanics* (Grundzüge der Strömungslehre) originated from introductory lectures which I have been giving for about 20 years at the University of Karlsruhe (TH). Here, I had the interesting task of introducing fluid mechanics to students after their preliminary examinations in a 4-h, one-semester lecture. The spectrum of the listeners was broad. It ranged from mechanical and chemical engineers to physicists, meteorologists, and mathematicians. This fact, together with the time available, determined the content and scope of the material presented. The aim was therefore not to present **everything** (that can be done in special lectures), but to choose a presentation that was as interesting as possible and easy for the students to grasp and apply.

A few words about the structure. Unlike most presentations of fluid mechanics, the momentum theorem is treated late. There are good reasons for this. Despite its simple formulation, it is and remains the most difficult theorem in fluid mechanics. The difficulty lies in the appropriate choice of the control space **and** the flow data used on the boundary. This is where a lot of knowledge goes in, which one has to gather **beforehand** when dealing with examples of fluid mechanics. We have made this experience again and again.

I have endeavored to follow a systematic structure. This is done by starting with the simplest and proceeding to the questions that arise in the numerous applications and are of great interest today. It is important, for example, to know from the beginning which and how many equations are available for the flow quantities. In some of the questions treated, one will feel certain attention to detail. This seems to be justified where students bring little information from other lectures. On the other hand, the beginner must have a thorough and detailed demonstration of the most important tools. That there will be compromises is clear to every lecturer.

Two-hour exercises are held parallel to the lectures. Without this own engagement of the listeners, one cannot master the material. Some of the assignments are included in the text. Here, as with the lecture material, the reader will have to reach for paper and pencil to absorb, process, and then apply the content. This effort is well worth it! I would be satisfied with the success of my many years of work if the reader could confirm this.

Karlsruhe, Germany   Jürgen Zierep

Contents

About the Authors

Jürgen Zierep held the chair for fluid mechanics at the Karlsruhe Institute of Technology KIT, has been Hon. Prof. at the BUAA (Beijing University of Aeronautics and Astronautics) since 1988, and is an internationally awarded and respected author of numerous technical books.

Karl Bühler teaches and conducts research at the University of Applied Sciences, Offenburg, in the Faculty of Mechanical and Process Engineering. His main fields of work are the fundamentals of frictional flows, boundary layer theory, rotating flows, vortex flows, solution properties of the Navier–Stokes equations, thermodynamics, convective heat transfer, instabilities in viscous, heat-conducting media, optical flow measurement technology, and numerical fluid mechanics.

List of Figures

Introduction, Overview and Basics

1

Abstract

Fluid mechanics deals with the motion processes in liquids and gases (so-called fluids). Instead of the term fluid mechanics, one often encounters the terms fluid mechanics, fluid dynamics, aerodynamics, etc. Fluid mechanics plays a major role in science and technology. The applications can be roughly divided into two groups:

1. Flow around bodies, e.g., motor vehicles, aircraft, buildings. Here, the flow field in the outer space is of interest, i.e. velocity, pressure, density and temperature in the vicinity of the body, which results in the force effect on the body.
2. Flow through pipes, ducts, machines and entire plants. Now the flow in the interior is of interest, e.g., of manifolds, diffusers and nozzles and the resulting pressure losses due to frictional influences.

The quantitative description of a flow is given by the flow quantities velocity, pressure, density and temperature, which are to be determined by the conservation laws and fluid properties.

Fluid mechanics deals with the **motion processes** in **liquids** and **gases** (so-called **fluids**). Instead of the term fluid mechanics, one often encounters the terms fluid dynamics, aerodynamics and others.

Fluid mechanics play a major role in science and technology. Roughly speaking, the applications can be divided into two different groups.

1. **Flow around** bodies, e.g., motor vehicles, aircraft, buildings. Here, the flow field in the **outer space** is of interest, i.e. velocity, pressure, density and temperature near and far from the body. This results, for example, in the force effect on the body from the flow around it.

2. **Flow through** pipes, ducts, machines and entire plants. Now the flow in the **interior** is of interest, e.g., of manifolds, diffusers and nozzles. Of importance here are frictional influences, which become noticeable through pressure losses.

In current technical problems, the two subtasks just discussed can of course also occur in combination. Numerous applications can be found in the fields of fluid mechanics, chemical engineering, aircraft construction, automotive engineering, building aerodynamics, meteorology, geophysics etc.

The **quantitative description of** a flow is given at any point (x, y, z) of the considered field at any time (t) by the quantities:

$$\text{Velocity } w = \left(u, v, w\right), \quad \text{Pressure } p, \quad \text{Density } \varrho, \quad \text{Temperature } T.$$

We assume the existence of these state variables as a function of $(x, y, z; t)$. We are thus in the realm of continuum mechanics. In total, we are dealing with 6 dependent and 4 independent variables. To determine the former 6 equations, the basic physical laws of fluid mechanics, are required. They are formulated in the form of conservation laws and are outlined in the following table:

	Physical statement	Number of equations	Type of equations
Conservation laws	**Continuity** (conservation of mass)	1	Scalar
	Equilibrium of forces (momentum theorem)	3	Vectorial
	Energy theorem (e.g., 1st law, Fourier's heat conduction equation[1] etc.)	1	Scalar
Fluid	**Equation of state** (thermodynamic linkage of p, ϱ, T)	1	Scalar

Compared with the mass point mechanics, which gets along with 3 equations for 3 velocity components, 6 equations are necessary here. To these differential or integral relations, initial conditions (t) and/or boundary conditions (x, y, z) are added to determine the solution of the problem, which may be **uniquely** determined, from the manifold of **possible** solutions. For the flow around and flow through problems given above, one can easily discuss these conditions. A general

---

[1] J.B.J. Fourier, 1768–1830.

solution of the basic equations of fluid mechanics encounters the greatest difficulties, since the associated differential equations are **nonlinear**. Therefore, one is often limited to so-called **similarity statements**, with which it is possible to transfer the flow data from one flow field to another. This leads to the important model laws which allow, for example, wind tunnel experiments to be converted to the large-scale model.

In the context of this presentation we will start with the simplest (hydrostatics). By increasing the number of independent and dependent variables, we will advance to the issues of interest in the applications. The following scheme explains from left to right our approach:

	Hydrostatic	Aerostatic	Hydrodynamic	Aerodynamic
p				
ϱ				
$w = (u, v, w)$				
Examples	Dormant liquid in the vessel	Dormant atmosphere	Moving fluid	Moving gas

The temperature T can be omitted here, since T can be determined from p and ϱ by the equation of state.

The historical development of fluid mechanics shows two different directions of work until about 1900.

1.1   Theoretical, Mainly Mathematical Fluid Mechanics

It is associated with the names Newton,[2] Euler,[3] Bernoulli,[4] D'Alembert,[5] Kirchhoff,[6] Helmholtz,[7] Rayleigh.[8] This is mainly the theoretical treatment of frictionless flows (so-called potential flows). With this, it was not possible, for example, to determine losses in flows quantitatively correctly in the case of flow-around and flow-through problems.

---

[2] I. Newton, 1643–1727.
[3] L. Euler, 1707–1783.
[4] D. Bernoulli, 1700–1782.
[5] J. D'Alembert, 1717–1783.
[6] G. Kirchhoff, 1824–1887.
[7] H. v. Helmholtz, 1821–1894.
[8] J.W. Rayleigh, 1842–1919.

1.2   Engineering Fluid Mechanics or Hydraulics

Decisive researchers were Hagen,[9] Poiseuille,[10] Reynolds.[11] Here it was a question of problems of measurement and their representation in friction flows, e.g., the laws for pipe flows.

Both directions were brought together in 1904 by Prandtl's[12] **boundary layer theory**. According to this theory, the cause of the frictional resistance of a body is to be found in the so-called boundary layer. This is a relatively thin layer close to the wall, in which the increase in velocity from zero at the wall to the value of the external flow occurs. Here the **adhesion condition** at the body surface is essential. If the body is moved, the flowing medium at the surface follows this movement. It sticks there! Figure 1.1 shows the particularly simple case of the longitudinally flowed plate—the prototype of a boundary layer. This Prandtl boundary layer concept has proved to be very fruitful. It leads to essential simplifications in the non-linear differential equations, so that a solution is possible.

If, in addition to frictional losses, heat transfer also plays a role, a **temperature boundary layer** occurs in addition to the **flow boundary layer** (Fig. 1.2). Both have their cause in completely analogous physical processes: friction and heat conduction.

Fig. 1.1 Flow boundary layer on the longitudinally flowed flat plate

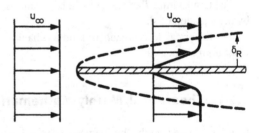

[9] G. Hagen, 1797–1884.

[10] J.L. Poiseuille, 1799–1869.

[11] O. Reynolds, 1842–1912.

[12] L. Prandtl, 1875–1953.

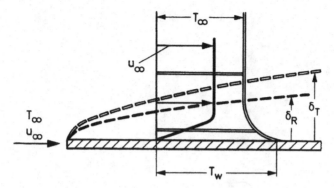

Fig. 1.2 Flow and temperature boundary layer on the flat plate

Properties of Fluids

2

Abstract

The molecular composition and the resulting microstructure is of central importance for the macroscopic properties of the fluid media. The molecular structure
is used to explain the physical states of gas, liquid and solid bodies. The resistance to changes in shape due to elasticity and viscosity is shown using the examples of solid elastic bodies and fluids and explained by the rheological sub-
stitute models.

Properties of Newtonian and non-Newtonian fluids are defined by the relationship between shear stress and strain rate. The behavior of dynamic and kinematic viscosity is shown as a function of temperature for fluids and gases.
With the gas-kinetic explanation of internal friction the kinematic viscosity is
attributed to molecular properties.

The equation of state of ideal gases is derived from the combination of
Boyle-Mariotte's and Gay-Lussac's law. The physical causes of surface and in-
terfacial tension and capillarity are considered in detail and explained in application examples.

2.1 Molecular Composition: Microstructure

For the understanding of the flow processes in fluids to be discussed later, it is important to compile the basic facts of their molecular composition. We are talking
about the microstructure.

© Springer Fachmedien Wiesbaden GmbH, part of Springer Nature 2022   7
J. Zierep, K. Bühler, *Principles of Fluid Mechanics*,
https://doi.org/10.1007/978-3-658-34812-0_2

Matter is made up of elementary components (molecules or atoms) whose diameter is of the following order of magnitude: $d \approx 10^{-10}$m. Now there are two facts that have a significant influence on the structure of these elements.

1. If the individual **particles** are relatively **far away** from each other, that is, if the density is sufficiently low, they are uninfluenced by each other and perform an irregular statistical movement due to their thermal energy (**Brownian movement for gases**[1]). For air under normal conditions, that is, atmospheric pressure and 20 °C, the following applies to this movement: Mean free path length $\ell \approx 10^{-7}$m, mean molecular velocity $\bar{c} \approx 500$m / s.

Addition:
Thermodynamics provides the dependence of the molecular velocity on the temperature. The internal energy of the molecule for each degree of freedom $=(1/2)kT$ with $k=$ Boltzmann constant.[2] With $f = 3$ as the number of degrees of freedom the energy theorem for the molecule of mass m:

$$\text{Internal energy} = \frac{3}{2}kT = \frac{1}{2}m\bar{c^2} = \text{Mean kinetic energy of translation.}$$

So $\bar{c^2} = \dfrac{3k}{m}T$, or in terms of size $\sqrt{\bar{c^2}} \approx \bar{c} \sim \sqrt{T}$.
This is a characteristic dependence on temperature, which occurs repeatedly later at many typical speeds (speed of sound, maximum speed).

2. If the **particles** are relative **close together**, that is, if the density is high enough, they influence each other. Intermolecular so-called van der Waals forces[3] act. Their extension reaches over a distance of about $10d \approx 10^{-9}$m. These attractive forces, which are electromagnetic forces by nature, can fix the particles mutually, for example, in a regular crystal lattice. Figure 2.1 gives a qualitative representation of the force exerted by a particle at zero point, on another at a distance r.

If the particles approach each other very closely, repulsion occurs instead of attraction. The inner structure of the particles plays a major role in this. However, this is not important for our considerations.

---

[1] R.Brown, 1773–1858 Botanist.
[2] L.Boltzmann, 1844–1906.
[3] J.D. van der Waals, 1837–1923.

Fig. 2.1 Intermolecular force exerted by a particle at zero point on another

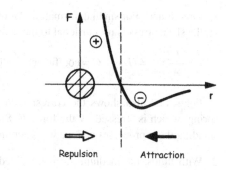

Fig. 2.2 Shear stress of a solid, elastic body and rheological model

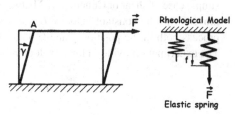

The interplay of the two facts **1.** and **2.** leads to the three **states of matter**. The following scheme explains this in a rough but satisfactory way. The density increases from left to right.

Gas	Liquid	Solid body
1. Predominates 2.	1. and 2. equal	2. Is predominant than 1.
Random motion	Random motion that is not unaffected by its neighbors takes place	Intermolecular forces bind particles to fixed points, e.g., in the crystal lattice

2.2 Resistance to Deformation (Elasticity, Viscosity)

There is a fundamental difference between **solid, elastic bodies** on the one hand, and **fluids** on the other. We explain this in the case of the claim to **shear stress** (**thrust**).

1. A **solid, elastic body** is stressed by a shearing force F. This case is sketched in Fig. 2.2. The angle γ is a measure for the deformation and A the area where the

force F acts. For small deformation, **Hooke's law**[4] applies, according to which, the shear stress is proportional to the deformation:

$$\frac{|F|}{A} = \tau = G \cdot \gamma, \quad G = \text{coefficient of sliding friction or sheer modulus.} \quad (2.1)$$

Figure 2.2 also shows the corresponding rheological model. It is the elastic spring, which is stressed by the force F. Such models are very useful for understanding these processes and we will encounter them often.

2. With fluids, the medium must be guided. Figure 2.3 explains the particularly simple case of shear or Couette flow[5] between two flat plates. The upper plate is moved with the constant velocity U, the lower one rests. We follow the flow process both in the position plane (x, y) as well as in the velocity plane (u, y). The experiment provides a linear velocity distribution

$$u = U\frac{y}{h} \quad (2.2)$$

in the plate gap. The adhesion condition for $y = 0$ and $y = h$ is apparently fulfilled. The connection between position and velocity plane leads to the equations

$$U = \frac{ds(t)}{dt}, \quad ds(t) = h d\gamma(t),$$

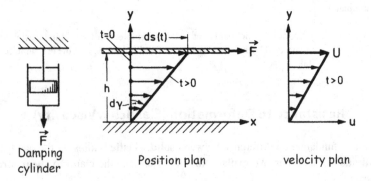

Damping cylinder   Position plan   velocity plan

Fig. 2.3 Couette flow. Position and velocity plane and rheological model for the Newtonian medium

---

[4] R. Hooke, 1635–1703.
[5] M. Couette, 1858–1943.

That is,

$$U = h\dot{\gamma}(t). \tag{2.3}$$

Let's define a **Newtonian fluid** through

$$\tau = \eta \frac{du}{dy}, \tag{2.4}$$

thus yield (2.2) and (2.3):

$$\tau = \eta \frac{du}{dy} = \eta \frac{U}{h} = \eta \dot{\gamma}(t). \tag{2.5}$$

For Newtonian fluids the shear stress is thus proportional to the deformation **velocity**. This is a fundamental difference to the elastic body. The rheological model here is a damping cylinder (Fig. 2.3).

The proportionality factor η in (2.4) and (2.5) means **dynamic viscosity**.

$$\nu = \frac{\eta}{\varrho} = \textbf{Kinematic viscosity}, \quad \varrho = \text{Density of medium}. \tag{2.6}$$

In the applications, the more general case is often found that

$$\tau = f(\dot{\gamma}) \tag{2.7}$$

f is called the flow function. Figure 2.4 contains some characteristic cases. Is f a **linear** function, we are dealing with Newtonian fluids (oil, water, air, etc.). The slope of the straight line is a direct measure of the dynamic viscosity. Nonlinear flow functions are used to describe **non-Newtonian fluids**. Examples are suspensions, polymers, oil paints etc. An interesting special case is the **Bingham medium**.[6] It behaves for $\tau < \tau_f$ like a solid, elastic body, but for $\tau > \tau_f$ like a Newtonian fluid. τ_f is called the yield stress. This is a model for the viscosity behavior of pulps and pastes. The rheological model (Fig. 2.5) contains a damping cylinder and parallel to it a block on a rough surface; upstream is an elastic spring whose deflection is limited. The block has static and sliding friction. The latter is neglected in the model under consideration. An interesting combination results when an elastic

---

[6] E.C. Bingham, 1878–1945.

Fig. 2.4 Various flow
functions. Newtonian,
non-Newtonian fluids,
Bingham medium

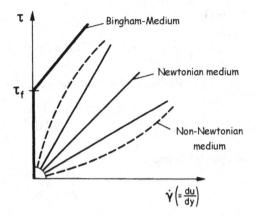

Fig. 2.5 Rheological model
of the Bingham medium

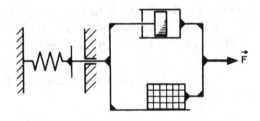

body shows viscous behavior. In this case one speaks of a **viscoelastic medium**.
Under short-term load, this medium behaves like an elastic body, whereas under
longer loads it behaves like a Newtonian fluid. The "jumping putty" is an example.
It is a kneadable, rubber-like medium. A ball is elastically reflected at the solid
wall. It melts under its own weight when lying down for a long time. The rheological model is an elastic spring and a damping cylinder connected in series
(Fig. 2.6).

Dimensions and units of measurement are required for the quantitative indica-
tion of the introduced physical quantities. We primarily use the international system (SI), but comparatively we also give the data in the old technical system.

The dynamic viscosity η is measured in

$$\frac{Ns}{m^2} = Pa \ s.$$

The kinematic viscosity $\nu = \eta/\varrho$ is measured in m^2/s.

The following table contains typical numerical values under normal conditions:

Fig. 2.6 Rheological model
for the viscoelastic medium

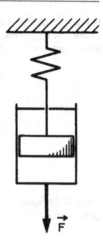

	$\eta \cdot 10^6$	$\nu \cdot 10^6$
	$\mathrm{Nm^{-2}s = Pa\,s}$	$\mathrm{m^2s^{-1}}$
Air	18.2	15.11
Water	1002.0	1.004
Silicone oil Bayer M 100	130,950.0	135.0

This immediately shows that η is characteristic for the force transmitted by the viscosity and not $\nu = \eta/\varrho$. (2.4) provides the statement for η

$$\eta = \frac{\tau}{\dfrac{\mathrm{d}u}{\mathrm{d}y}},$$

that is, η is a measure of the force per unit area ($=\tau$), which is required to reach the velocity gradient ($=\mathrm{d}u/\mathrm{d}y$). The kinematic viscosity is obtained by dividing by the density, whereby the order of magnitude can be changed considerably.

Viscosity is **temperature-dependent** for fluids; With increasing temperature it decreases in liquids and increases in gases (Fig. 2.7). This fact becomes understandable from the microstructure shown above. Viscosity is a macroscopic effect caused by the molecular, that is, microscopic impulse exchange of the individual fluid particles. This view leads in the next section to the gas-kinetic explanation of internal friction. In gases, the molecular velocity increases with increasing temperature and thus the momentum transmitted during the impact, which leads to an increase in viscosity in this model. In liquids the intermolecular forces play a decisive role. Here, increasing temperature loosens the mutual bond, the particles become more easily displaceable and the viscosity decreases.

Fig. 2.7 Viscosity as a
function of temperature for
liquids and gases

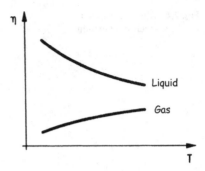

Fig. 2.8 Dynamic viscosity
of the sulfur melt

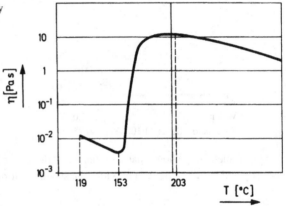

An interesting special case is the sulfur melt (Fig. 2.8). In a certain temperature
range, this substance behaves like a liquid, when the temperature increases it behaves like a gas and when the temperature increases further it behaves like a liquid
again. This is obviously connected with the transformation of the crystalline structure of this substance. Interesting is the **analogy** between **internal friction and
heat conduction**. These are molecular transport processes that proceed in a similar
way.

We can put a very simple heat conduction problem aside for the Couette flow
discussed above (Fig. 2.9). The heat flow $\dot{Q}$, as the amount of heat transferred per
time, can be represented using the Fourier approach as follows:

$$\dot{Q} = -\lambda A \frac{dT}{dy}. \tag{2.8}$$

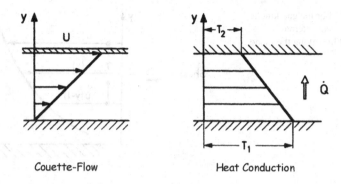

Couette-Flow   Heat Conduction

Fig. 2.9 Analogy of internal friction and heat conduction

λ is the thermal conductivity and A the transmitting surface. For the specific heat flow $\dot{q}$ therefore applies

$$\frac{\dot{Q}}{A} = \dot{q} = -\lambda \frac{dT}{dy}. \tag{2.9}$$

Thus the analogy between internal friction and heat conduction is

$$\tau = \eta \frac{du}{dy} \quad \leftrightarrow \quad \dot{q} = -\lambda \frac{dT}{dy}. \tag{2.10}$$

λ and η are molecular exchange variables for heat and momentum. For later it is important that a characteristic number can be derived from both quantities. It is named after Prandtl:

$$\text{Prandtl number} = \text{Pr} = \frac{\eta c_p}{\lambda}. \tag{2.11}$$

c_p is the specific heat at constant pressure.

2.3 Gas Kinetic Explanation of Internal Friction

Two statements are at issue here. On the one hand we want to derive Newton's shear stress approach (2.4) for gases and on the other hand we want to trace η or ν back to known microscopic values. For the derivation we use gas-kinetic considerations, where macroscopic and microscopic considerations are considered simulta-

Fig. 2.10 For the gas-kinetic
explanation of internal
friction. Flow along the plane
wall

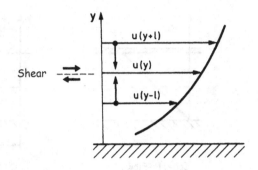

neously. We investigate the flow along a plane wall (Fig. 2.10). $u(y)$ is the averaged
speed profile as we register it macroscopically, for example, with the naked eye.
Microscopically, on the other hand, the gas particles execute a random motion. The
resulting **impulse exchange** of the different layers leads to an interlocking, that is,
to internal friction. Figure 2.10 illustrates how the shear occurs in the layer y. ℓ
denotes the mean free path length. Particles, which originate from the level $(y + \ell)$
accelerate the particles present at y. Particles coming from below $(y - \ell)$ decelerate
them accordingly. This causes a shear or shear stress in the level y. This force effect
is now calculated.

The mass of the impacting particle is m, then the average momentum of a par-
ticle coming from above is $|i_o|$ $= mu(y + \ell)$; the same applies to a particle that
comes from below, $|i_u|$ $= mu(y - \ell)$.

To determine the total transmitted momentum, we have to count how many
particles pass through the unit area per unit time, n being the number of particles
per cm³. Due to the equal distribution, $n/3$ particles move in 1 cm³ in x- or y- or
z-direction, that is, half of them, so $n/6$ particles, in $(+x)$- or $(-x)$-, ... -direction.
So kick $(n/6)\bar{c}$ particles per second by the unit area (Fig. 2.11). The decisive
factor here is the average microscopic molecular speed $\bar{c}$ of the particles.
Figure 2.11 shows that only those molecules that are located in a cuboid with the
height $\bar{c} \cdot 1$s can pass through the unit area in the unit of time.

Thus for the impulse transmitted to the layer y per unit of time and area = force
per area = shear stress.

$$
\tau = \frac{n}{6}\bar{c}mu\left(y+\ell\right) - \frac{n}{6}\bar{c}mu\left(y-\ell\right)
$$

$$
= \frac{nm}{6}\bar{c}\left[u\left(y\right)+\ell\frac{du}{dy}+\cdots-u\left(y\right)+\ell\frac{du}{dy}+\cdots\right]. \tag{2.12}
$$

Fig. 2.11 Determination of
the number of impacting
particles

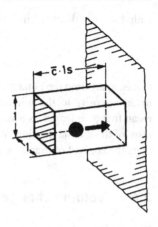

During development, we only take into account the ℓ linear terms, then

$$\tau = \frac{nm\bar{c}\ell}{3}\frac{du}{dy} = \frac{\varrho\bar{c}\ell}{3}\frac{du}{dy},$$ (2.13)

$$\frac{\eta}{\varrho} = \nu = \frac{\bar{c}\ell}{3}.$$ (2.14)

Thus, the Newtonian shear stress approach for gases is derived and at the same time ν accesses the previously introduced microscopic determinants $\bar{c}$ and ℓ back to the original location. Because of $\bar{c} \sim \sqrt{T}$ further follows that for gases the viscosity ν increases with temperature. From the specifications for air: $\bar{c} \approx 500\text{m/s}$, $\ell \approx 10^{-7}\text{m}$ follows with (2.14) $\nu \approx 15 \cdot 10^{-6}\text{m}^2/\text{s}$ which is consistent with the one in the table in Sect. 2.2 and is in good agreement.

With a **dimensional view**, you can also go to the presentation of the kinematic viscosity ν. It is assumed that ν can depend solely on the kinematic microscopic determinants ℓ and $\bar{c}$

$$\nu = f\left(\ell, \bar{c}\right).$$

A power approach

$$\nu = A\bar{c}^m\ell^n$$

with A = const delivers immediately

$$v = A\bar{c}\ell,$$

that is, apart from the numerical factor A, again (2.14). This reduction of the kinematic viscosity to the microscopic quantities: Average molecular velocity and mean free path length, is a very plausible result, which clearly shows the mechanism of the internal friction. We will come back to similar considerations later on (e.g., the concept of Prandtl's mixing length) repeatedly.

2.4   Volume Change and Equation of State for Gases

We recall two elementary basic laws for so-called ideal gases.

1. The **Boyle-Mariotte** law[7] for **isothermal** processess states:

$$pV = p_0 V_0 = \text{const}, \quad t = \text{const}, t \text{ in } {}^\circ\text{C}. \tag{2.15}$$

2. The **Gay-Lussac** law[8] for **isobaric** processes is:

$$V = V_0 \left(1 + \beta t \right), \quad p = \text{const}, \beta = \frac{1}{273\,^\circ\text{C}}. \tag{2.16}$$

t herein refers to the Celsius temperature. In Fig. 2.12 are different processes (isochoric $- \varrho$ = const, isobaric $- p$ = const, isothermal $- t$ = const, isentropic $- s$ = const) in the (p, V)-plane are entered.

Any change of state p_1, V_1, $t_1 = 0 \rightarrow p$, V, t can always consist of the two elementary processes considered above (Fig. 2.13). An **isothermal** process leads from p_1, V_1, $t_1 = 0$ to p, V_2, $t_2 = 0$:

$$p_1 V_1 = p V_2. \tag{2.17}$$

A subsequent **isobaric** process leads p, V_2, $t_2 = 0$ in p, V, T over with

[7] R. Boyle, 1627–1691; E. Mariotte, 1620–1684.
[8] J.L. Gay Lussac, 1778–1850.

Fig. 2.12 Thermodynamic state changes in the (p, V)-plane

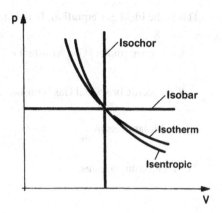

Fig. 2.13 Composition of an isothermal and an isobaric process

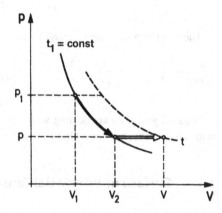

$$V = V_2 (1 + \beta t).$$ (2.18)

If we use (2.17) here, then

$$pV = pV_2 (1 + \beta t) = p_1 V_1 (1 + \beta t) = \beta p_1 V_1 \left(\frac{1}{\beta} + t \right) = \beta p_1 V_1 T,$$

thus in specific sizes

$$pv = \frac{p}{\varrho} = \frac{R}{m} T = R_i T.$$ (2.19)

This is the **ideal gas equation**. In this, mean:[9]

$$R = \text{general}\,(\text{molar})\,\text{Gas}\quad \text{constant} = 8.314 \cdot 10^3 \frac{m^2 g}{s^2 \text{molK}} = 8.314 \frac{J}{\text{molK}},$$

$$R_i = \text{specific or special Gas}\quad \text{constant in} \frac{m^2}{s^2 K} = \frac{J}{\text{kgK}},$$

$$m = \text{Molar mass in} \frac{g}{\text{mol}}.$$

The relationship applies

$$\frac{R}{m} = c_p - c_v = R_i. \tag{2.20}$$

The most important representatives are

Gas	O_2	N_2	H_2	Air
m in g / mol	32	28.016	2.016	29

For **real** gases, we are dealing with more general equations of state in contrast to the above considerations. One example is the so-called van-der-Waals-equation.

2.5   Surface or Interfacial Tension and Capillarity

So far we have only considered **a homogeneous** medium. Now we are dealing with the immediate surroundings of the interface between **two** media of different density. Such surfaces play an important role in fluid mechanics. If two immiscible liquids of different densities touch each other, we speak of an internal interface. A free surface is present when a liquid and a gas are adjacent to each other. **Inside** of the liquid, the intermolecular attractive forces on a particle cancel each other out on average (Fig. 2.14). A spherically symmetrical force field is present here. At the **surface** a resultant occurs which is directed inwards, as the gas particles above the surface do not exert any intermolecular forces. The thickness of this surface layer is comparable to the range of action of the intermolecular forces ($\approx 10d \approx 10^{-9}$m). This resultant of the intermolecular forces is in balance with the other forces (gravity, pressure force). In anticipation of later developments, this means that a modi-

---

[9] J.P.Joule, 1818–1889.

Fig. 2.14 Intermolecular forces inside and on the surface of a liquid

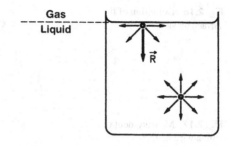

Fig. 2.15 The formation of minimal surfaces

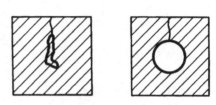

fied hydrostatic pressure distribution occurs near the surface. Work must be done against this resultant if a particle is to be moved from the interior of the liquid to the surface, that is, the molecules on the surface have a higher energy than particles inside. So there is energy in every liquid surface. To keep this additional energy low, nature uses as few particles as possible to form the surface. This leads to the formation of so-called **minimal surfaces**. The following example illustrates this clearly (Fig. 2.15). A rectangular wire frame is filled with a soap skin. A closed thread of yarn is placed on top. If the soap skin is pierced in the loop, the loop will spring open to form a circle. As is well known, the circle has the largest area for a given circumference. The remaining area is therefore the minimum area under the given boundary conditions.

The effort to make the surface as small as possible leads to **stress state in the surface**. The **surface tension** σ is defined as the force per unit length of the boundary that keeps the surface in balance (Fig. 2.16). The surface tension σ depends essentially on the two media that are adjacent to each other on the surface. Moreover, it decreases with increasing temperature. The explanation is similar to that for viscosity. The intermolecular bonds decrease with increasing temperature. The resultant in Fig. 2.14 decreases, and thus the surface energy decreases.

Media	σ in Nm^{-1}
Water/air	0.071
Oil/air	0.025–0.030

Fig. 2.16 Definition of the
surface tension

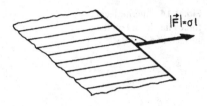

$|\vec{F}|=\sigma\,l$

Fig. 2.17 Measurement of σ
with a wire hanger

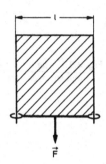

Media	σ in Nm^{-1}
Mercury/air	0.46

(at 20 °C)

The measurement of σ can be carried out with a wire bow, on which a bar is attached in a sliding manner (Fig. 2.17). Since two surfaces are formed, the following applies

$$|\,F\,| = 2\sigma\ell$$

Instead of the

$$\text{Surface tension } \sigma = \frac{\text{Attacking force}|\,F\,|\text{ on the edge}}{\text{Length of the edge}\,\ell} \qquad (2.21)$$

is often referred to as the

$$\text{Specific surface energy}\,\varepsilon = \frac{\text{Increase in energy } \Delta E}{\text{Increase in surface area}\,\Delta A} \qquad (2.22)$$

is used. It is

$$\varepsilon = \sigma. \qquad (2.23)$$

Fig. 2.18 On the equality of surface tension and specific surface energy

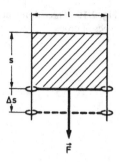

In order to prove this statement, we move the bridge in Fig. 2.18 by Δs. The surface increase achieved is $\Delta A = 2 \Delta s \ell$. With (2.22), the following therefore applies to the energy required

$$\Delta E = 2\varepsilon \Delta s \ell. \tag{2.24}$$

On the other hand, if we move the bridge with the attacking force $|F| = 2\sigma\ell$ about Δs, it perform the work

$$\Delta W = 2\sigma \Delta s \ell. \tag{2.25}$$

Since both energies (2.24) and (2.25) are equally large (energy balance!), it follows

$$\Delta W = 2\sigma \Delta s \ell = 2\varepsilon \Delta s \ell = \Delta E,$$

that is, it applies (2.23).

With this relationship it is easy to estimate the energy contained in liquid surfaces. This is especially important if surfaces are constantly being formed anew, which is the case with atomization, for example. This last observation leads us away from **flat surfaces** treated so far to **curved surfaces**, which are of great interest in the applications. Examples are a water drop, a liquid bubble or a gas drop (Fig. 2.19). In all three cases the surface tension strives to compress the drop or bubble. This leads to an increase in pressure inside. If gravity is neglected, this leads to an equilibrium between the pressure force and the force resulting from the surface tension.

Fig. 2.19 Three different
types of drops

1.Liquid drop:

2. Liquid bubble:

3. Gas drops:

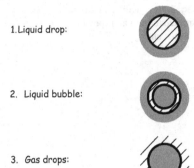

Fig. 2.20 Equilibrium
consideration for the
spherical liquid drop

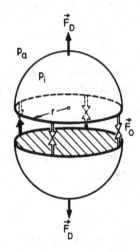

Particularly simple is the associated **equilibrium consideration** for the **spherical liquid drops** (Fig. 2.20). If we cut the drop at the equator, we get a resulting **surface tension force**

$$|F| = 2\pi r\sigma. \qquad (2.26)$$

The resulting **pressure force** points in the vertical direction and has the same size as if the equatorial plane would be impinged with Δp (Fig. 2.21):

Fig. 2.21 To derive the
pressure force

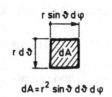

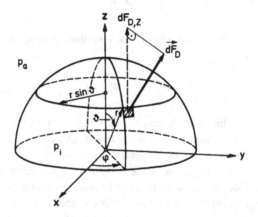

$$| \boldsymbol{F}_{\mathrm{D}} | = \Delta p \pi r^2 = \left(p_{\mathrm{i}} - p_{\mathrm{a}} \right) \pi r^2. \tag{2.27}$$

From the balance of the two forces follows

$$\Delta p = p_{\mathrm{i}} - p_{\mathrm{a}} = \frac{2\sigma}{r}. \tag{2.28}$$

For the bubble there is an additional factor of two on the right side of (2.28), because two surfaces are formed. This can result in a considerable **overpressure** in the drop or in the bubble. In the case of fog droplets, $r = 10^{-6}\mathrm{m} = 10^{-3}\mathrm{mm}$, we receive for example:

$$\Delta p = \frac{2 \cdot 7.1 \cdot 10^{-2}}{10^{-6}} \frac{\mathrm{N}}{\mathrm{m}^2} = 1.42 \cdot 10^5 \frac{\mathrm{N}}{\mathrm{m}^2} = 1.42 \mathrm{bar}.$$

We want the detailed derivation of the **pressure force** (2.27) because the result is of general interest. In Fig. 2.21, the interface element is specified. For reasons of symmetry, only the z-component of the pressure force must be considered:

The external pressure p_a results in the following proportion for the area element

$$-p_a dA \cos\vartheta = -p_a r^2 \sin\vartheta \cos\vartheta \, d\vartheta \, d\varphi.$$

Integration across the hemisphere delivers

$$-\int_{\varphi=0}^{2\pi} \int_{\vartheta=0}^{\pi/2} p_a r^2 \sin\vartheta \cos\vartheta \, d\vartheta \, d\varphi = -p_a \pi r^2.$$

Taking into account the contribution of p_i on the equatorial surface $=p_i \pi r^2$ the total is

$$F_{D,z} = \left(p_i - p_a\right)\pi r^2 = \Delta p \pi r^2.$$

This fact is generally valid, that is, at curved surfaces only the projection into the respective plane is important. The projection is in the direction in which the component of the pressure force is sought.

We now come to the equilibrium consideration with an arbitrarily curved surface. We cut out a rectangular surface element. Figure 2.22 contains all the terms relating to geometry and forces. r_1 and r_2 are the radii of curvature of the intersection curves of the surface with two planes perpendicular to each other.

Fig. 2.22 Equilibrium consideration for an arbitrarily curved surface

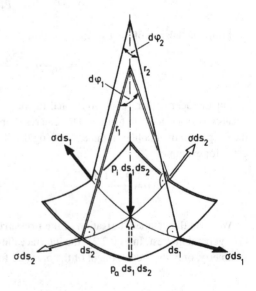

Fig. 2.23 Balance of forces
for the case of Fig. 2.22

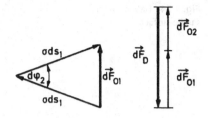

Notice that r_1 and r_2 are afflicted with signs. They have opposite signs if the centers of curvature are on different sides of the surface. The following relationships apply

$$ds_1 = r_1 d\varphi_1, \quad ds_2 = r_2 d\varphi_2. \tag{2.29}$$

The resultant of the surface tensions applied must keep the pressure force in equilibrium (Fig. 2.23):

$$d\boldsymbol{F}_{01} + d\boldsymbol{F}_{02} + d\boldsymbol{F}_{D} = 0. \tag{2.30}$$

The following applies here for the amounts

$$dF_{D} = \left(p_i - p_a\right)ds_1 ds_2 = \Delta p ds_1 ds_2,$$
$$dF_{01} = \sigma ds_1 d\varphi_2 = \frac{\sigma}{r_2}ds_1 ds_2, \tag{2.31}$$
$$dF_{02} = \sigma ds_2 d\varphi_1 = \frac{\sigma}{r_1}ds_1 ds_2.$$

The equilibrium condition provides

$$\Delta p = p_i - p_a = \sigma\left(\frac{1}{r_1} + \frac{1}{r_2}\right). \tag{2.32}$$

As **special cases** we would like to point out:

1. Bubble drop: $r_1 = r_2 = r$, $\Delta p = \dfrac{2\sigma}{r}$
2. Spherical bubble: $\Delta p = \dfrac{4\sigma}{r}$
3. Cylindrical surface: $r_1 = r$, $r_2 \to \infty$, $\Delta p = \dfrac{\sigma}{r}$

Fig. 2.24 Balance on a
saddle surface

A sign discussion is of interest. We start with $r_1 > 0$ and $r_2 > 0$. Both centers of curvature are on the same side of the surface. We hold on $r_1 > 0$ and let r_2 vary from positive values to infinity to negative values. Then the curvature in the both directions one and two is opposite. It is the behavior of a saddle surface (Fig. 2.24). In such a case, it is quite possible $\Delta p = p_i - p_a = 0$ namely when the surface tension forces are in equilibrium with each other. With suitable wire hangers, such saddle surfaces can be easily produced from soap skin. Both sides of the surface are under atmospheric pressure, and thus the following applies to the radii of curvature

$$\frac{1}{r_1} + \frac{1}{r_2} = 0. \tag{2.33}$$

In mathematics, this defines the so-called minimal surfaces. The condition for this is that the average surface curvature

$$H = \frac{1}{2}\left(\frac{1}{r_1} + \frac{1}{r_2} \right) = 0$$

is what agrees with our condition (2.33).

Finally some elementary consequences. If two bubbles are in contact with each other, the convex side is in the larger bubble, because the smaller one has the higher pressure (Fig. 2.25). If the two bubbles are connected by a pipe, the small one in-

Fig. 2.25 Two bubbles in contact

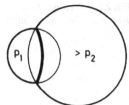

Fig. 2.26 Pressure equalization between two bubbles of different sizes

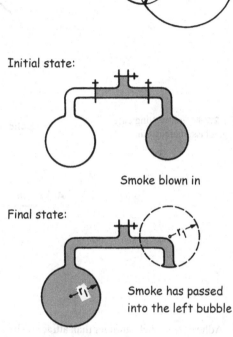

flates the large one. In the final state both surfaces have the same curvature. In Fig. 2.26 the corresponding experiment is sketched. In the initial state there are two bubbles of different sizes. The smaller one on the right contains smoke. When the connection between the two is made, the smoke flows to the left. The right bubble decreases in size. In the final state, a piece of surface remains on the right, which has the same curvature as the bubble on the left.

We come to the discussion of the **capillarity**. The term derives from the raising or lowering of the liquid level in a capillary. Now it is a matter of three media, for example, gas-liquid-solid body. Besides the intermolecular forces of the liquid, the attractive forces of the wall (=adhesion) also play a role. The weight of the particle itself (gravity) can be disregarded in this context.

Two extreme cases are of importance.

Fig. 2.27 Wetting of the
wall (e.g., glass, water, air)

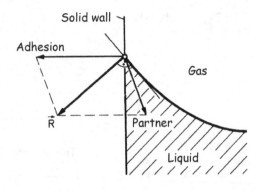

Fig. 2.28 Non-wetting case
(e.g., glass, mercury, air)

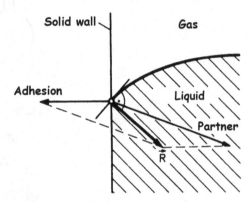

1. Adhesion is much stronger than attraction by neighboring liquid particles. At this
   case the wall is wetted. The liquid is attracted to the wall and rises to the top
   (Fig. 2.27). The resultant of adhesion and intermolecular forces is perpendicular
   to the surface of the liquid (neglecting gravity!). This must be so, because otherwise there would be a component in the direction of the surface, which would
   lead to a displacement of the particles at the surface and thus to a movement. The
   sketched case corresponds, for example, to combination: glass, water and air.

2. If the attraction of the liquid partners is much greater than the adhesion, the non-
   wetting case is present. The liquid sinks down the wall (Fig. 2.28). This corresponds, for example, to the combination: glass, mercury, and air.

Fig. 2.29 Balance in
capillary uplift

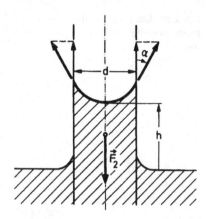

Of importance is the calculation of the capillary **rise** (or fall). We want to use
different methods here, all of which play a role in the applications.

1. We use the **cutting principle of the mechanics** and cut the liquid column raised
   in the capillary free (Fig. 2.29). There must then be a balance between the surface tension force F_1 and the weight force of the lifted liquid column F_2. In
   equilibrium with the designations of Fig. 2.29

$$F_1 = 2\pi r \sigma \cos\alpha = F_2 = \pi r^2 h \varrho g,$$

that is,

$$h = \frac{2\sigma \cos\alpha}{r\varrho g} = \frac{4\sigma \cos\alpha}{d g \varrho}. \qquad (2.34)$$

Adhesion enters through the contact angle α. When fully wetted, $\alpha = 0$, and the
following applies

$$h = \frac{4\sigma}{d g \varrho}, \qquad (2.35)$$

which provides an easy way to measure the surface tension σ.

2. We consider the special case of complete wetting and explain another possibility (Fig. 2.30). A pressure smaller than atmospheric pressure is created in the
   liquid under the free surface

Fig. 2.30 Calculation of the
pressures during capillary
rise

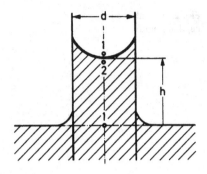

$$\Delta p = p_\mathrm{i} - p_2 = \frac{2\sigma}{r} = \frac{4\sigma}{d}.$$

The associated suction force, which results from projection of the surface into the capillary cross section, supports the liquid column. The following applies to the suction force: $\Delta p \pi r^2 = 2\pi\sigma r$ while the weight is $\pi r^2 h \varrho g$. If we put both expressions in the same way, we get (2.35).

3. We use the consistency condition for the pressure in point 2 (Fig. 2.30). At this point we anticipate the representation for hydrostatic pressure. The pressure in 2 can be calculated in two ways: As hydrostatic pressure $p_2 = p_1 - g\varrho h$ and as capillary pressure $p_2 = p_1 - 2\sigma/r$. By equating you get (2.35) again. From the last display you can immediately see that the height of capillary rise is limited. For $p_2 = 0$ is $h_{max} = p_1/(\varrho g)$ or rather $r_{min} = 2\sigma/p_1$. Strictly speaking, it would be required that the pressure of the liquid must not fall below the vapor pressure of the liquid at point 2. The first represents the well known capillary height of the hydrostatic head, the second gives, within the framework of the model used here, an estimate of the minimum capillary radius.

With $\sigma = 7.1 \cdot 10^{-2}\mathrm{N/m}$ and $p_1 = 10^5\mathrm{N/m^2}$ we get $d_{min} \cong 3 \cdot 10^{-3}\mathrm{mm}$. For the capillary heights in the cylindrical tube (2.35) follows

$$h_{\mathrm{H_2O}} = \frac{28.8}{d}\mathrm{mm}, \ \# h_{\mathrm{Hg}}\# = \frac{13.8}{d}\mathrm{mm}.$$

Fig. 2.31 Capillary rise
between two adjacent plates

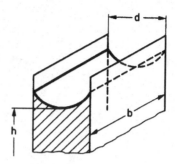

In this is d to be entered in mm.

If we consider the uplift of a liquid between two adjacent vertical plates (Fig. 2.31), then for the vertical component of the surface tension force

$$F_1 = 2b\sigma \cos\alpha$$

and for the weight of the lifted liquid

$$F_2 = hbd\varrho g,$$

so

$$h = \frac{2\sigma \cos\alpha}{dg\varrho}.$$

The capillary height is therefore half as large as with the cylindrical capillaries. This is immediately obvious, since the effective pressure is also only half as high.

Hydro- and Aerostatics

3

Abstract

Hydro- and aerostatics deals with the state variables in the absence of motion. The pressure p is proved to be a scalar quantity by a force balance for Newtonian fluids. Fluid pressure in force fields is derived using the basic hydrostatic equation for the gravity and centrifugal force fields. A stratified medium is present in the atmosphere. The pressure and density distributions are derived and discussed for the troposphere and stratosphere.

Technically important applications are considered with the calculation of the pressure force on plane vessel walls. The hydrostatic paradox is explained. Hydrostatic buoyancy and the associated Archimedes' principle are derived and applied to the examples. With some experience, forces on curved surfaces can be determined using Archimedes' principle. Considerations of the stability of floating bodies round off this chapter.

3.1   Liquid Pressure p

In this chapter we deal with the state variables in the absence of motion. We first catch up with the proof that the **pressure** is a **direction-independent quantity**, i.e., a **scalar** quantity. We consider a mass element with the depth dz (Fig. 3.1) of a Newtonian fluid in a motionless equilibrium state. Shear stresses occur in the intersecting surfaces only when there is motion. In our case, only normal forces are present there. We label the pressure with an index: p_x, p_y, p_z, depending on which surface it acts on. The weight enters as a volume force. These are the relations

© Springer Fachmedien Wiesbaden GmbH, part of Springer Nature 2022   35
J. Zierep, K. Bühler, *Principles of Fluid Mechanics*,
https://doi.org/10.1007/978-3-658-34812-0_3

Fig. 3.1 Equilibrium of
forces at the stationary mass
element

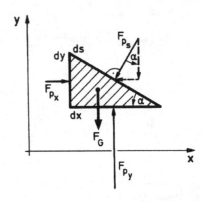

$$dx = ds\cos\alpha, \quad dy = ds\sin\alpha.$$

The equilibrium of forces is

$$\sum_\alpha F_\alpha = 0,$$

i.e., the sum of all acting forces is zero. In components this means

$$\sum F_x = p_x dydz - p_s \sin\alpha\, ds dz = \left(p_x - p_s\right) dydz = 0,$$
$$p_s = p_x.$$
(3.1)

$$\sum F_y = p_y dxdz - p_s \cos\alpha\, ds dz - \frac{1}{2}dxdydz\varrho g$$

$$= \left(p_y - p_s - \frac{1}{2}\varrho g dy\right) dxdz = 0,$$
(3.2)

$$p_s = p_y - \frac{1}{2}\varrho g dy.$$

If we contract the mass element to a point, then (3.1) and (3.2) merge into

$$p_s = p_x = p_y.$$
(3.3)

We can therefore omit the indices. Thus the **fluid pressure** p is recognized as a
scalar quantity.

3.2   Fluid Pressure in Force Fields

We still consider a medium at rest, cut out an infinitesimal cuboid (Fig. 3.2) and formulate the equilibrium condition. We divide the occurring forces into two classes: **mass forces** and **surface forces**. By definition, the first group includes all forces acting on the mass of the element (gravity, centrifugal force, electric and magnetic forces, etc.). The second group contains all forces acting on the surface (pressure force, friction forces, etc.). The latter forces are caused by normal and tangential stresses on the surface. In the special case considered here, only static pressure occurs.

The mass force per unit mass is

$$f = \left\{ f_x, f_y, f_z \right\}. \tag{3.4}$$

The x-component of the equilibrium condition is

$$p \mathrm{d}y \mathrm{d}z - \left(p + \frac{\partial p}{\partial x} \mathrm{d}x \right) \mathrm{d}y \mathrm{d}z + f_x \mathrm{d}m = 0, \quad \text{also,}$$

$$-\frac{\partial p}{\partial x} \mathrm{d}x \mathrm{d}y \mathrm{d}z + f_x \mathrm{d}m = 0.$$

With the mass element $\mathrm{d}m = \varrho \mathrm{d}x \mathrm{d}y \mathrm{d}z$ follows for all three components

$$\frac{1}{\varrho}\frac{\partial p}{\partial x} = f_x, \quad \frac{1}{\varrho}\frac{\partial p}{\partial y} = f_y, \quad \frac{1}{\varrho}\frac{\partial p}{\partial z} = f_z \tag{3.5a}$$

Fig. 3.2 Equilibrium consideration for the cuboid

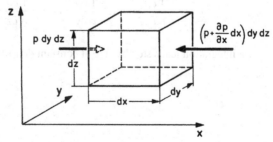

Fig. 3.3 On centrifugal force

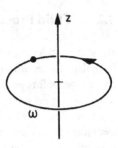

or, summarized in vector form,

$$\frac{1}{\varrho} grad\, p = f. \tag{3.5b}$$

We discuss special cases that are important for fluid mechanics. As mass forces we consider gravity

$$f_s = \{0,0,-g\} \tag{3.6}$$

as well as the centrifugal force that occurs when rotating with constant angular velocity ω about the *z-axis* (Fig. 3.3):

$$f_z = \{\omega^2 x, \omega^2 y, 0\}. \tag{3.7}$$

Strictly speaking, this example no longer falls within the field of statics. If the fluid rotates with constant angular velocity like a rigid body, the above equations can still be used. (3.6) and (3.7) result with (3.5b) in the differential equation system.

$$\frac{1}{\varrho}\frac{\partial p}{\partial x} = \omega^2 x, \quad \frac{1}{\varrho}\frac{\partial p}{\partial y} = \omega^2 y, \quad \frac{1}{\varrho}\frac{\partial p}{\partial z} = -g. \tag{3.8}$$

The integration leads step by step at constant density ϱ to the result

Fig. 3.4 Isobars with rotation of the fluid around the *z-axis*

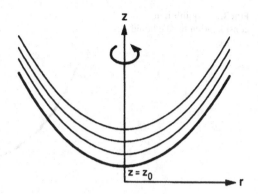

$$p(x,y,z)=\frac{1}{2}\varrho\omega^2 x^2+f(y,z) \quad \rightarrow \quad \frac{\partial f(y,z)}{\partial y}=\varrho\omega^2 y,$$

$$f(y,z)=\frac{1}{2}\varrho\omega^2 y^2+h(z) \quad \rightarrow \quad \frac{dh}{dz}=-g\varrho,$$

$$h(z)=-g\varrho z+\text{const}, \tag{3.9}$$

$$p(x,y,z)=\frac{1}{2}\varrho\omega^2\left(x^2+y^2\right)-g\varrho z+\text{const}.$$

This pressure distribution can be most easily discussed in terms of the isobaric surfaces $p=$ const. These are rotational paraboloids, which all diverge by translation along the axis of rotation (Fig. 3.4):

$$z-z_0=\frac{\omega^2}{2g}\left(x^2+y^2\right)=\frac{\omega^2}{2g}r^2. \tag{3.10}$$

Especially this representation is valid for the liquid surface, because there the pressure is equal to the constant atmospheric pressure. It is also very easy to arrive at this result in the following way: at the liquid surface, the resultant of gravity and centrifugal force must be perpendicular to the surface (Fig. 3.5). This leads to the increase

$$\frac{dz}{dr}=\frac{\omega^2 r}{g},$$

i.e., (3.10) applies:

Fig. 3.5 Equilibrium consideration for the liquid surface

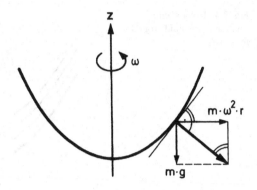

Fig. 3.6 Pressure distribution of liquids at rest in the gravitational field

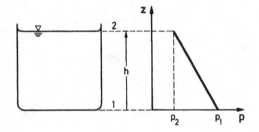

$$z - z_0 = \frac{\omega^2}{2g} r^2.$$

We are now concerned with the pressure **distribution** in liquids and gases in the **gravitational field.**

In a fluid at rest ($\varrho = $ const) we obtain from (3.9) for the pressure difference $p_1 - p_2$ at the height difference h (Fig. 3.6)

$$\Delta p = p_1 - p_2 = g \varrho h. \qquad (3.11)$$

Thus, pressure increases **linearly** with depth in liquids. The **units of** pressure are[1]

---

[1] B. Pascal, 1623–1662.

$$1\mathrm{Pa}\,(\mathrm{Pascal}) = 1\frac{\mathrm{N}}{\mathrm{m}^2} = 1\frac{\mathrm{kg}}{\mathrm{ms}^2},$$

$$1\mathrm{bar} = 10^5\,\mathrm{Pa} = 10^5\,\frac{\mathrm{N}}{\mathrm{m}^2}.$$

Older units are

- Technical atmosphere:

$$1\ \mathrm{at} = 1\frac{\mathrm{kp}}{\mathrm{cm}^2} = 10\ \mathrm{mWS} = 0.981\ \mathrm{bar}$$

- Physical atmosphere:[2]

$$1\mathrm{atm} = 760\mathrm{Torr} = 76\mathrm{cmHg} = 1.033\mathrm{at} = 1.013\mathrm{bar} = 1013\mathrm{mbar}$$

The following relationships apply

$$1\ \mathrm{Torr} = \frac{1}{760}\ \mathrm{atm} = 133.3\ \mathrm{Pa},$$

$$1\ \mathrm{mmHg} = 13.6\ \mathrm{mmWS}.$$

The older definitions assume either the pressure of a 10 m high water column (WS) (=1 at) or a 76 cm high mercury column (=1 atm).

The pressure measurement (p_1) can thus be traced back to a length measurement (h) (Fig. 3.7). This is the principle of the barometer. The maximum height of rise is

Fig. 3.7 Principle of the barometer

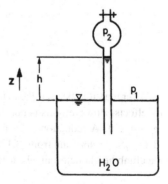

---

[2] E. Torricelli, 1608–1647.

Fig. 3.8 The barometer
considering capillarity effects

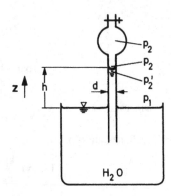

obtained for $p_2 = 0$. If p_1 is the atmospheric pressure and the vapor pressure is neglected, $h_{max} = 10 \text{mWS}$.

Of interest is the consideration of the barometer taking into account capillarity effects (Fig. 3.8). Hydrostatics provides

$$p_1 - p_2' = g\varrho h. \tag{3.12}$$

The surface tension leads to a pressure difference at complete wetting.

$$p_2 - p_2' = \frac{4\sigma}{d} \tag{3.13}$$

Subtraction results in

$$p_1 - p_2 = g\varrho h - \frac{4\sigma}{d},$$

i.e.,

$$h = \frac{p_1 - p_2}{g\varrho} + \frac{4\sigma}{dg\varrho}. \tag{3.14}$$

After that, it looks as if the maximum rise height could be increased by capillarity effects. However, this is not the case. The only to note here is that after (3.13) $p_2 - p_2' \geq 0$. A reduction of p_2 has its limit where $p_2' = 0$. Then it follows from (3.13): $p_{2min} = 4\sigma/d$ and from (3.12): $p_1 = g\varrho h_{max}$. This is in complete agreement with the climb height determined earlier.

Let us now examine the pressure distribution in a **stratified medium** with density $\varrho = \varrho(z)$. There are many applications, e.g., in liquids or also in the atmosphere. We will deal with gas layers in the following. The basic hydrostatic equation for pressure (3.8)

$$\frac{dp}{dz} = -g\varrho$$

yields with the ideal gas equation (2.19)

$$\frac{p}{\varrho} = R_i T$$

the equation of determination

$$\frac{dp}{p} = -\frac{g}{R_i}\frac{dz}{T}. \tag{3.15}$$

An integration is only possible if $T = T(z)$ is given. This is an additional statement, which follows from thermodynamic considerations (energy balance!). The integration becomes particularly simple for an **isothermal** gas layer. Equation (3.15) results with the initial values $p(z_0) = p_0$, $\varrho(z_0) = \varrho_0$:

$$p = p_0 \exp\left(-\frac{g}{R_i T_0}(z - z_0)\right), \quad \varrho = \varrho_0 \exp\left(-\frac{g}{R_i T_0}(z - z_0)\right). \tag{3.16}$$

In the case of a fluid of constant density, the dependence of the pressure on the height is linear. Here, on the other hand, it is an exponential curve (Fig. 3.9).

Fig. 3.9 Pressure distribution in an isothermal gas layer

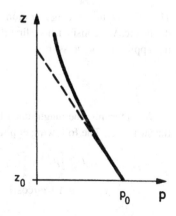

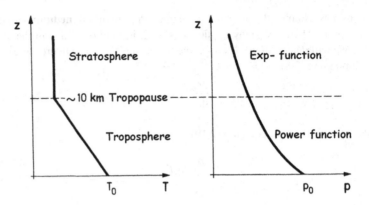

Fig. 3.10 Temperature and pressure in the atmosphere

In the atmosphere, in the troposphere and in the stratosphere, the temperature can be well approximated by a straight line and by a constant, respectively (Fig. 3.10). An appropriate integration as above leads to a power function in the troposphere, while an exponential function in the stratosphere. Both pressure functions merge into each other in the tropopause in a continuous and continuously differentiable manner. The latter follows immediately from the basic hydrostatic Eq. (3.5b) with (3.6), because the density is continuous in the transition. In meteorology, these relations (barometric altitude formulas) are used very frequently, for e.g., in ascent evaluation.

3.3   Pressure Force on Flat Container Walls

The subject under consideration is important for dimensioning vessels, tanks, dams, etc. We consider here first the **plane** inclined walls (Fig. 3.11). The following applies to the pressure

$$p = p_1 + g\varrho z.$$

We determine the magnitude of the force $|\boldsymbol{F}| = F$ transmitted from the fluid to the surface A. The following applies to $dF = pdA$, i.e., in total

$$F = \int_A p dA = \int_A (p_1 + g\varrho z) dA = p_1 A + g\varrho \cos\alpha \int_A \ell dA$$
$$= p_1 A + g\varrho \cos\alpha \ell_s A = p_1 A + g\varrho z_s A = p_s A.$$

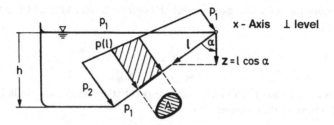

Fig. 3.11 Pressure force on flat, inclined walls

Here the ℓ_s centroid coordinate of A, is defined by

$$\int_A \ell dA = \ell_s A. \tag{3.17}$$

The force exerted by the fluid is thus equal to the pressure at the focus of the surface multiplied by the surface area. If the constant pressure p_1 prevails outside, the resulting force is

$$F_{Res} = g\varrho z_s A. \tag{3.18}$$

This is an obvious result. In determining the force, the lower pressures above the center of gravity apparently cancel out with the over pressures below the center of gravity. This is due to the **linear** distribution of pressures. It is different for the moments needed to determine the point of application of this force. The overpressures below the centre of gravity have a larger lever arm than the under pressures. Consequently, the point of application of the force is always **below** the center of gravity. We perform the calculation only for constant external pressure p_1, i.e., for the resultant (3.18). The reader can treat the general case immediately. The moment equilibrium with respect to the x-axis is:

$$F_{Res}\ell_m = g\varrho z_s A\ell_m = \int_A (p - p_1)\ell dA = \int_A g\varrho z\ell dA = g\varrho\cos\alpha\int_A \ell^2 dA = g\varrho\cos\alpha J_x.$$

J_x denotes the area moment of inertia of A with respect to the x-axis. Thus, the contact point of the force F_{Res} is:

$$\ell_m = \frac{J_x}{A\ell_s}. \tag{3.19}$$

We move the reference axis parallel through the center of gravity. Steiner's theorem[3] is

$$J_x = J_s + A\ell_s^2$$

with J_s as the area moment of inertia with respect to the gravity axis parallel to the x-axis. So from (3.19) becomes

$$\ell_m - \ell_s = \frac{J_s}{A\ell_s} > 0, \qquad (3.20)$$

which shows that the point of application of the force is below the centre of gravity. This deviation can be considerable. In the special case of a rectangular, flat, vertical wall (Fig. 3.12), the result is, for example

$$\ell_s = \frac{h}{2}, \quad \ell_m = \frac{2}{3}h.$$

The above considerations immediately yield the explanation of the so-called hydrostatic paradox (Fig. 3.13). According to (3.18), the resulting force on the bottom surface of the various containers depends only on A, z_s and ϱ, but not on the **shape of the container**. The force exerted on the bottom surface is the same in all sketched cases, although the weight of the liquid contained in the vessels is different.

Fig. 3.12 The centre of gravity (ℓ_s) and point of application of the resulting force (ℓ_m) for the rectangular, plane, vertical wall

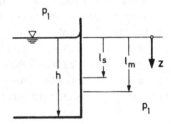

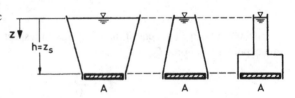

Fig. 3.13 The hydrostatic paradox

---

[3] J. Steiner, 1796–1863.

3.4 Hydrostatic Buoyancy. Pressure Force on Curved Surfaces

We consider a body completely immersed in a liquid (Fig. 3.14). Due to the hydrostatic pressure distribution, the pressure at the bottom of the body is greater than at the top. This results in a vertically directed force, the buoyancy. The observation is first carried out for one body element:

$$dF_z = p_2 dA_2 \cos\beta - p_1 dA_1 \cos\alpha = (p_2 - p_1) dA$$
$$= g\varrho_{Fl} h dA = g\varrho_{Fl} dV.$$

Integration delivers

$$F_z = g\varrho_{Fl} V, \tag{3.21}$$

i.e., the buoyancy is equal to the weight of the displaced fluid (Archimedes' principle[4]). This statement can be used directly to determine ϱ_{Fluid} or ϱ_{Body}. For the body weight G applies

$$G = g\varrho_{Body} V,$$

so with (3.21)

$$\frac{G}{F_z} = \frac{\varrho_{Body}}{\varrho_{Fluid}}.$$

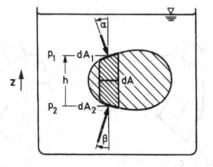 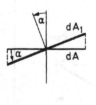

Fig. 3.14 The hydrostatic buoyancy

---

[4]Archimedes of Syracuse, 287–212 BC.

If one of the specific weights is known, the other can be determined if G and F have been measured.

Archimedes' principle also applies to partially immersed bodies (Fig. 3.15). To prove this, a section is made through the body along the surface of the liquid with constant pressure p_1 along the section. The pressure integral over the non-submerged part of the body (I) gives the value zero, since the pressure is constant. What remains is the volume of the immersed part (=V), and thus (3.21) also applies here.

Archimedes' principle can be applied very simply to determine the forces on **curved** surfaces, since the integration over the arbitrary body shape has already been carried out here once and for all. We illustrate this with a simple example (Fig. 3.16). A cone of revolution points horizontally into a vessel filled with fluid. We determine the components F_x and F_z of the force acting on the cone. We cut the cone free and obtain the specified force distribution. We record the linear pressure distribution on the cone plan—once positive and once negative. Thus we have on the

Fig. 3.15 Archimedes' principle for the partially immersed body

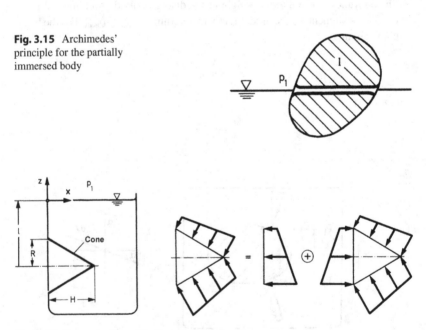

Fig. 3.16 Determination of the force on curved surfaces

one hand a completely immersed body (=Cone) to which we can apply Archimedes' principle, and on the other hand a **plane** immersed surface. This gives us

$$F_x = p_s \pi R^2 = \left(p_1 + g\varrho_{Fl}\ell\right)\pi R^2,$$

$$F_z = \frac{1}{3}g\varrho_{Fl}\pi R^2 H.$$

Another application concerns the **floating** of a body in a liquid. In this case, it is the **balance of forces** between buoyancy and weight of the body. Figure 3.17 illustrates this. S_K is the body's centre of gravity, while S_v denotes the centre of gravity of the displaced fluid. In general $S_K \neq S_v$, since the masses may be unevenly distributed or the body may be only partially submerged. Therefore, the question of the **stability** of this equilibrium state arises. For this purpose, we bring the body slightly out of the equilibrium position and consider the restoring moment of the buoyancy force.

In Fig. 3.18 a **stable** case is considered. The buoyancy provides a back-rotating moment. Stability obviously exists whenever the **metacenter** M is above the body's center of gravity. M is the intersection of the line of action of the lift with the vertical axis of the body. Figure 3.19 illustrates an **unstable** case. M lies below S_K. These illustrative considerations are plausible and very simple. However, the quantitative determination, e.g., of the oscillations around the equilibrium position, requires some effort.

Fig. 3.17 Balance of forces
during swimming

Immersed
floating body

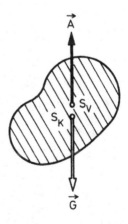

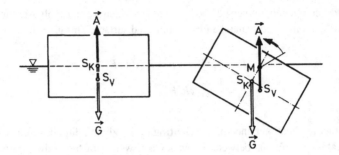

Fig. 3.18 Stable balance during swimming

Fig. 3.19 Unstable
equilibrium during swimming

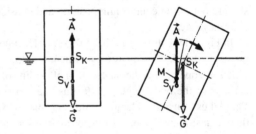

Hydro- and Aerodynamics

4

Abstract

The basic concepts of steady and unsteady flows are derived according to Lagrange for the mass- or particle-fixed consideration and Euler for the stationary consideration. Streamline theory is considered with conservation laws for mass, momentum, and energy. The gas dynamic considerations include the speed of sound, the flow in the Laval nozzle and the vertical compression shock.

Frictionless, plane flows form the basis of the potential theory, which is explained by numerous examples.

The flows with friction are treated with the momentum theorem. Dimensionless ratios illustrate the fundamental influence of friction in laminar-turbulent transition. The laminar and turbulent pipe flow leads over to the basic properties of turbulent flows: Prandtl's mixing length, Reynolds apparent shear stress, and logarithmic wall law.

The general form of the Navier-Stokes equations, boundary layer theory, the energy theorem, and similarity considerations with dimensional analysis conclude in this chapter.

4.1   Flow Filament Theory

4.1.1   Basic Concepts

The following must be determined for a moving medium

© Springer Fachmedien Wiesbaden GmbH, part of Springer Nature 2022   51
J. Zierep, K. Bühler, *Principles of Fluid Mechanics*,
https://doi.org/10.1007/978-3-658-34812-0_4

Fig. 4.1 Motion along the particle path

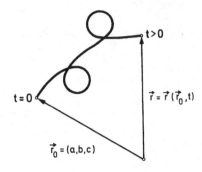

$$w = (u,v,w),\ p,\ \varrho, T. \tag{4.1}$$

For this purpose—as already emphasized in the introduction—six equations are available. In the following, special cases of these equations for the study of motion are examined. The totality of the quantities (4.1) in the considered space and time domain describes a **flow field**. This field is called **stationary** if all quantities (4.1) are only functions of the spatial coordinates. On the other hand, the field is called **unsteady** if time appears as an additional variable.

There are two different ways of describing flow fields.

1. **Lagrangian method**[1] **(mass- or particle-fixed observation)**

Here, the individual particle is tracked as it moves through space. The respective position of the particle is a function of the initial position

$$r_0 = (a,b,c)$$

and the time t. The particle trajectory (Fig. 4.1) is thus written in the form

$$r = r(r_0, t). \tag{4.2}$$

For the velocity w and the acceleration b, the following material or substantial derivatives result (Fig. 4.2):

$$w = \lim_{\Delta t \to 0} \frac{\Delta r}{\Delta t} = \lim_{\Delta t \to 0} \frac{r(t+\Delta t) - r(t)}{\Delta t} = \left(\frac{\partial r}{\partial t}\right)_{a,b,c} = \frac{dr}{dt}, \tag{4.3}$$

---

[1] J.L. Lagrange, 1736–1812.

Fig. 4.2 Formation of the substantial derivative

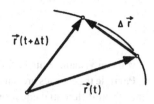

$$b = \left(\frac{\partial^2 r}{\partial t^2} \right)_{a,b,c} = \frac{d^2 r}{dt^2}. \tag{4.4}$$

The index a,b,c means that the derivation is carried out at a fixed initial position, that is, for one and the same particle. However, the measurements required for this are difficult to realize. One would have to let the measuring instrument fly along, so to speak. On the other hand, this description is well suited for quantities which are firmly connected with the particle in question. For example, the vorticity introduced below is such a quantity. All conservation laws (mass, momentum and energy) are best formulated this way.

2. **Euler's method (stationary observation)**

Here we consider the change in the flow quantities at a fixed point in space as the individual particles pass by. This corresponds to the procedure when measuring with a fixed measuring device. Both representations are related in a simple way. For a particle property $f(x, y, z, t)$

the chain rule gives

$$\frac{df}{dt} = \frac{\partial f}{\partial t} + \frac{\partial f}{\partial x}\frac{dx}{dt} + \frac{\partial f}{\partial y}\frac{dy}{dt} + \frac{\partial f}{\partial z}\frac{dz}{dt}$$

$$= \frac{\partial f}{\partial t} + \frac{\partial f}{\partial x}u + \frac{\partial f}{\partial y}v + \frac{\partial f}{\partial z}w$$

$$= \frac{\partial f}{\partial t} + w\,\mathrm{grad}\,f. \tag{4.5}$$

Here on the left side is the substantial change, while on the right side the local change occurs in the first place. The difference between the two is formed by the **convective** expression $w\,\mathrm{grad}f$. It describes in a simple way the influence of the velocity field.

Using the example $f = T$, that is,

$$\frac{\mathrm{d}T}{\mathrm{d}t} = \frac{\partial T}{\partial t} + w\,\mathrm{grad}\,T,$$

you can visualize this very easily.

Particle trajectories are curves through which the particles pass in the course of time. Their differential equation results from (4.2) and (4.3) to

$$\frac{\mathrm{d}\boldsymbol{r}}{\mathrm{d}t} = w,$$

that is,

$$\frac{\mathrm{d}x}{\mathrm{d}t} = u(x,y,z,t), \quad \frac{\mathrm{d}y}{\mathrm{d}t} = v(x,y,z,t), \quad \frac{\mathrm{d}z}{\mathrm{d}t} = w(x,y,z,t). \tag{4.6}$$

If the velocity w is known, the particle trajectories are obtained by integration.

Streamlines are curves that fit the velocity field at any fixed time. They represent an instantaneous picture of the velocity field (Fig. 4.3). At a later time, the shape of the streamlines may be quite different. The differential equation in the (x, y)-plane is (Fig. 4.4)

$$\frac{\mathrm{d}y}{\mathrm{d}x} = \frac{v(x,y,z,t)}{u(x,y,z,t)}.$$

t plays the role of a parameter here. In general, the differential equations can be summarized in the following relation:

$$\mathrm{d}x : \mathrm{d}y : \mathrm{d}z = u(x,y,z,t) : v(x,y,z,t) : w(x,y,z,t). \tag{4.7}$$

Fig. 4.3 Streamlines as instantaneous image of the velocity field

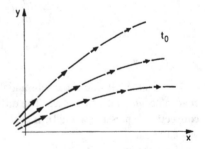

Fig. 4.4 On the differential
equation of the streamlines

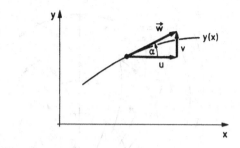

Fig. 4.5 Stationary flow
around the circular cylinder

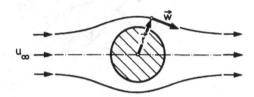

In **stationary** flows, the particle trajectories coincide with the streamlines. In
(4.7) then no more time dependence occurs. In the case of **unsteady** flows, on the
other hand, the two curve systems generally differ. We explain this problem with a
simple example.

We consider the flow around a cylinder at rest with the incident flow u_∞
(Fig. 4.5). The observer is located on the cylinder. The flow is **stationary**. The
particle trajectories coincide with the streamlines. We now change the reference
frame and move the observer along with the incident flow. The cylinder then moves
from right to left with velocity $-u_\infty$. Now we are dealing with an **unsteady** flow.
The cylinder pushes the medium in front of it as it moves, pushing it to the side and
finally leaving it behind. Figure 4.6 shows the instantaneous images of the streamlines at two different times and a particle trajectory. Figure 4.7 shows some particle
trajectories for different initial positions. If the particle is far away from the cylinder in the transverse direction, it performs an almost circular evasive motion. If the
particle is approached to the cylinder, it performs a looping motion, the horizontal
extent of which increases as it approaches the axis. The discussion of these particle
motions is very interesting and provides many insights into fluid mechanics. The
quantitative calculation of particle trajectories uses the potential theory developed
below. It is a characteristic of the example discussed that the unsteady flow can be
made a steady flow merely by changing the reference frame.

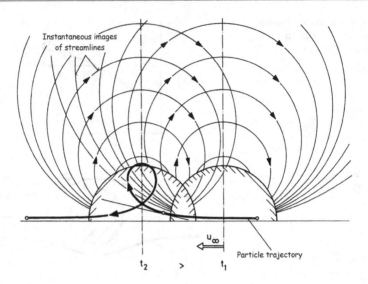

Fig. 4.6 Unsteady flow during movement of the cylinder. Instantaneous images of the streamlines and particle trajectory

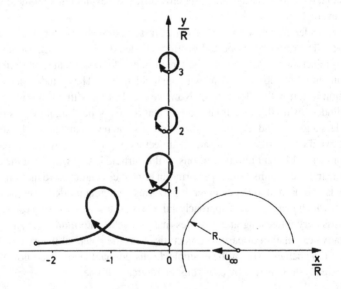

Fig. 4.7 Different particle trajectories during cylinder motion

4.1.2   Basic Equations of the Flow Filament Theory

Here we consider **frictionless** flows which, moreover, are supposed to be **stationary with** a few exceptions. For the following, the concept of the **streamline** is decisive. We start from a streamline $1 \rightarrow 2$ (Fig. 4.8). In 1 we consider the cross-sectional area A_1. Through each edge point of A_1 we draw another streamline. These envelop a **streamtube**. The **flow filament** represents an abstraction. Here we restrict ourselves to the immediate vicinity of the streamline in such a way that the changes of all state variables in the transverse direction are much smaller than in the longitudinal direction. Thus, in each cross-section of the flow filament there is only one value each for velocity c, pressure p, density ϱ and temperature T. These quantities then depend only on the arc length s and, if applicable, on the time t. This makes it a one-dimensional process, the treatment of which is much simpler than the general case. This **flow filament theory** is an important tool for fluid mechanics. However, if one wants to treat examples from applications with it, one must check carefully whether the conditions for the idealization made are fulfilled. In particular, it must be checked whether the changes in the state variables in the transverse direction are really very much smaller than in the longitudinal direction.

1. Continuity equation (constancy of mass flow)

The sheath of the flow filament consists of streamlines (Fig. 4.8). Nothing passes through it. Therefore, the mass passing through the cross-section per unit time is

$$\dot{m} = \varrho_1 c_1 A_1 = \varrho_2 c_2 A_2 = \text{constant}$$

or generally

$$\dot{m} = \varrho c A = \text{constant}. \tag{4.8}$$

Fig. 4.8 Definition of the flow filament

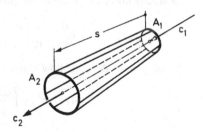

Fig. 4.9 Equilibrium of forces in the direction of the flow filament

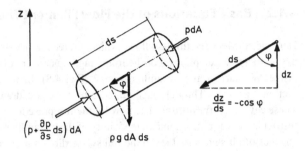

2a.Balance of forces in the direction of the flow filament

In the following consideration on the infinitesimal flow filament element, the change in cross-section can be neglected. It gives higher order members in the differentials. We apply Newton's fundamental law (Fig. 4.9):

$$\text{Mass} \times \text{Acceleration} = \text{Sum of the applied forces.} \qquad (4.9)$$

$$\text{Mass} = \mathrm{d}m = \varrho \mathrm{d}A \mathrm{d}s,$$

Herein is

$$\text{Acceleration} = \frac{\mathrm{d}c}{\mathrm{d}t} = \frac{\partial c}{\partial t} + \frac{\partial c}{\partial s}\frac{\mathrm{d}s}{\mathrm{d}t} = \frac{\partial c}{\partial t} + c\frac{\partial c}{\partial s},$$

Applied forces = pressure + weight

$$= -\frac{\partial p}{\partial s}\mathrm{d}A\,\mathrm{d}s + \varrho g\,\mathrm{d}A\,\mathrm{d}s\cos\varphi = -\left(\frac{\partial p}{\partial s} + \varrho g\frac{\partial z}{\partial s}\right)\mathrm{d}A\,\mathrm{d}s.$$

(4.9) gives **Euler's equation** for the flow filament

$$\frac{\mathrm{d}c}{\mathrm{d}t} = \frac{\partial c}{\partial t} + c\frac{\partial c}{\partial s} = -\frac{1}{\varrho}\frac{\partial p}{\partial s} - g\frac{\partial z}{\partial s}. \qquad (4.10)$$

for **steady** flows all quantities are only functions of s:

$$c\frac{\mathrm{d}c}{\mathrm{d}s} = \frac{\mathrm{d}}{\mathrm{d}s}\left(\frac{c^2}{2}\right) = -\frac{1}{\varrho}\frac{\mathrm{d}p}{\mathrm{d}s} - g\frac{\mathrm{d}z}{\mathrm{d}s}. \qquad (4.11)$$

An integration along the flow filament from $1 \to 2$ results in

$$\frac{1}{2}\left(c_2^2 - c_1^2\right) + \int_{p_1}^{p_2}\frac{\mathrm{d}p}{\varrho} + g\left(z_2 - z_1\right) = 0. \qquad (4.12a)$$

If we consider the final state (2) as variable, then

$$\frac{c^2}{2} + \int^p \frac{dp}{\varrho} + gz = \text{constant.} \tag{4.12b}$$

The constant summarizes the three terms on the left in the initial state (1). It is the same for all points of the flow filament, but can vary from flow filament to flow filament. (4.12a, b) is called Bernoulli equation and provides an important relation between velocity and pressure.

For **unsteady** flows, the Bernoulli equation on the left contains the additional element

$$\int_1^2 \frac{\partial c}{\partial t} \, ds. \tag{4.13}$$

The integration is to be carried out there with fixed t along the streamline of $1 \to 2$. This expression must often be estimated and compared with the terms appearing in (4.12a), in order to be sure that one may calculate stationary.

In Bernoulli's equation (4.12b) each member has the dimension of one energy per mass. Nevertheless, this is **not** the energy theorem, but an integral of the equation of motion. In continuum mechanics this is essential. For the evaluation of the integral

$$\int_{p_1}^{p_2} \frac{dp}{\varrho} \tag{4.14}$$

in (4.12a) it must be known from an energy balance what change of state of $1 \to 2$ takes place. We come back to this with (4.23a, 4.23b, 4.23c, 4.23d).

2b. Equilibrium of forces perpendicular to the flow filament

Flow filaments can exert forces on each other. Figure 4.10 sketches the case of a curved flow filament. The following results in sequence

$$\text{Mass} = dm = \varrho dA \, dn,$$

$$\text{Acceleration in normal direction} = \frac{dc_n}{dt} = -\frac{c^2}{r}.$$

r is the local radius of curvature of the orbit. The minus sign occurs because the acceleration points to the center of curvature.

$$\text{Applied forces} = -\frac{\partial p}{\partial n} dA \, dn + \varrho g \, dA \, dn \sin\varphi$$

$$= -\left(\frac{\partial p}{\partial n} + \varrho g \frac{\partial z}{\partial n} \right) dA \, dn.$$

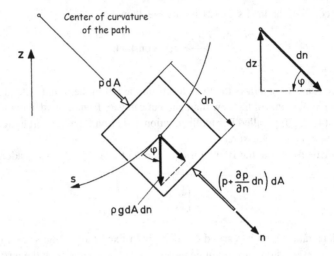

Fig. 4.10 Balance of forces perpendicular to the flow filament

Thus follows

$$\frac{c^2}{r} = -\frac{1}{\varrho}\frac{\partial p}{\partial n} + g\frac{\partial z}{\partial n}. \tag{4.15}$$

Without gravity, we have an equilibrium between centrifugal force and pressure force:

$$\frac{c^2}{r} = -\frac{1}{\varrho}\frac{\partial p}{\partial n}, \tag{4.16}$$

that is, in radial direction pressure increases. The force resulting from this pressure increase keeps the centrifugal force in balance.

3. **Energy theorem for steady-state flow filament theory**

We summarize the internal energy (e) and the kinetic energy ($1/2\,c^2$) per unit mass:

$$e + \frac{1}{2}c^2. \tag{4.17}$$

This energy flow in the flow filament is thus

$$\dot{E} = \left(e + \frac{1}{2}c^2\right)\dot{m}. \tag{4.18}$$

In both cross sections 1 and 2 we obtain:

$$\dot{E}_1 = \left(e_1 + \frac{1}{2}c_1^2 \right) \varrho_1 c_1 A_1 = \left(e_1 + \frac{1}{2}c_1^2 \right) \dot{m}, \qquad (4.19a)$$

$$\dot{E}_2 = \left(e_2 + \frac{1}{2}c_2^2 \right) \varrho_2 c_2 A_2 = \left(e_2 + \frac{1}{2}c_2^2 \right) \dot{m}. \qquad (4.19b)$$

The cause for the change of the energy flow of $1 \rightarrow 2$ is given by the power of the applied forces as well as by the mechanical power and the power of the heat flow. If we denote the work supplied to the unit of mass by w_t and the heat supplied by q, we get

$$\dot{E}_2 - \dot{E}_1 = p_1 A_1 c_1 - p_2 A_2 c_2 + g\left(z_1 - z_2 \right) \dot{m} + w_t \cdot \dot{m} + q \cdot \dot{m}. \qquad (4.20)$$

With (4.19a, b) follows

$$e_2 + \frac{p_2}{\varrho_2} + \frac{1}{2}c_2^2 + gz_2 = e_1 + \frac{p_1}{\varrho_1} + \frac{1}{2}c_1^2 + gz_1 + w_t + q \qquad (4.21a)$$

or with the enthalpy $h = e + p/\varrho$

$$h_2 + \frac{1}{2}c_2^2 + gz_2 = h_1 + \frac{1}{2}c_1^2 + gz_1 + w_t + q. \qquad (4.21b)$$

If we take the final state (2) as variable again, then

$$h + \frac{1}{2}c^2 + gz - w_t - q = \text{constant}. \qquad (4.22a)$$

This equation has a remarkable relationship with Bernoulli's equation (4.12b), with which, however, it agrees only in special cases. We will come back to it. The energy eq. (4.21a) can be represented for the special case of incompressible flow without heat supply as:

$$\frac{p_1}{\varrho} + \frac{1}{2}c_1^2 + g \cdot z_1 + w_t = \frac{p_2}{\varrho} + \frac{1}{2}c_2^2 + g \cdot z_2 + \frac{1}{\varrho}\Delta p_v. \qquad (4.22b)$$

The pressure loss due to friction is associated with an increase in internal energy $e_2 - e_1 = \Delta p v / \varrho$.

Here the technical work is $w_t > 0$ at energy input (pump) and $w_t < 0$ at energy output (turbine).

We summarize the result. Along the flow filament (s) we have the following three nonlinear equations for the variables c, p and ϱ:

$$\dot{m} = \varrho c A = \text{constant},\qquad\qquad(4.22c)$$

$$\frac{1}{2}c^2 + \int^p \frac{dp}{\varrho} + gz = \text{constant}.\qquad\qquad(4.22d)$$

$$\frac{1}{2}c^2 + h + gz - w_t - q = \text{constant}.\qquad\qquad(4.22e)$$

A and q are herein considered to be known. The enthalpy h is due to thermodynamics to p and ϱ. The equilibrium of forces normal to the flow filament provides the pressure change $\partial p / \partial n$, if with the above system $c(s)$ and were determined $\varrho(s)$.

Instead of the three basic equations (4.22c), (4.22d) and (4.22e), one can of course proceed to combinations of them. If, for example, one subtracts (4.22d) and (4.22e), one obtains

$$h - \int^p \frac{dp}{\varrho} - w_t - q = \text{constant}.\qquad\qquad(4.22f)$$

This corresponds in differential form to

$$dh - \frac{dp}{\varrho} = dw_t + dq,\qquad\qquad(4.22g)$$

that is, the first law of thermodynamics comes, which then takes the place of the law of energy. If no work or heat is added or removed and no friction occurs, then (4.12b) is for $\varrho = $ constant identical to (4.22a). This is the important special case in which Bernoulli's equation coincides with the energy theorem. The difference between the two equations becomes essential only when energy components occur that are not included in the equation of motion. Examples of this are: Supply and removal of work and heat, heat conduction processes, radiation components.

For simple changes of state, we can easily determine the integral occurring in (4.12b).

Isobar:

$$p = \text{constant},\qquad \int_1^2 \frac{dp}{\varrho} = 0.\qquad\qquad(4.23a)$$

Here comes the energy theorem of mass point mechanics:

$$\text{kinetic} + \text{potential energy} = \text{constant}.$$

Isochor:

$$\varrho = \text{constant}, \quad \int_1^2 \frac{dp}{\varrho} = \frac{p_2 - p_1}{\varrho} = \frac{\Delta p}{\varrho}. \tag{4.23b}$$

Isothermal: $T = \text{constant}.$
The ideal gas equation leads to

$$\frac{dp}{\varrho} = R_i T \frac{d\varrho}{\varrho}, \quad \int_1^2 \frac{dp}{\varrho} = R_i T \ln \frac{\varrho_2}{\varrho_1}. \tag{4.23c}$$

Isentrop: the reversible adiabatic $\dfrac{p}{p_1} = \left(\dfrac{\varrho}{\varrho_1}\right)^\kappa$,

$$\int_1^2 \frac{dp}{\varrho} = \frac{p_1^{1/\kappa}}{\varrho_1} \int_1^2 \frac{dp}{p^{1/\kappa}} = -\frac{\kappa}{\kappa-1} \frac{p_1}{\varrho_1} \left[1 - \left(\frac{p_2}{p_1}\right)^{\frac{\kappa-1}{\kappa}}\right]. \tag{4.23d}$$

4.1.3 Flow Filament Theory in Detail

In this section we will cover in detail, among other things, a wide range of examples. This will help you understand many typical details of fluid mechanics, which we will return later.

4.1.3.1 Movement on Concentric Circular Paths (Vortices)

The movement takes place in a horizontal plane. Then we can disregard gravity. Because of the rotational symmetry all quantities depend only on r and not on the polar coordinate angle φ. In radial direction is valid (4.16):

$$\frac{c^2}{r} = \frac{1}{\varrho} \frac{dp}{dr} \tag{4.24}$$

and in the circumferential direction (4.12b):

$$\frac{1}{2} c^2 + \int^r \frac{dp}{\varrho} = f(r). \tag{4.25}$$

Here we consider that the total energy $f(r)$ can in principle depend on r *in* the flow field. If we assume below $f(r) \equiv$ constant, we restrict ourselves to isoenergetic flows. With this additional assumption, the Bernoulli equation also relates the states on different streamlines. The continuity equation does not give any statement here, since $A = A(r)$ is not known. (4.24) and (4.25) are two equations for c, p and ϱ. Thermodynamics provides the missing condition with a prescription for the type of change of state.

For isoenergetic flows, (4.25) gives with (4.24)

$$0 = c \frac{dc}{dr} + \frac{1}{\varrho} \frac{dp}{dr} = c \frac{dc}{dr} + \frac{c^2}{r},$$

that is,

$$\frac{dc}{dr} = -\frac{c}{r}$$

with the solution ($r = r_1, c = c_1$)

$$c = \frac{c_1 r_1}{r}. \tag{4.26}$$

This is a hyperbolic velocity distribution $c \sim 1/r$. This is referred to as a **potential vortex**. For the calculation of the pressure we restrict ourselves to isochoric processes. (4.24) gives

$$\frac{1}{\varrho} \frac{dp}{dr} = \frac{c^2}{r} = \frac{c_1^2 r_1^2}{r^3}$$

with the solution ($r = r_1, p = p_1$)

$$p = p_1 + \frac{\varrho}{2} c_1^2 r_1^2 \left(\frac{1}{r_1^2} - \frac{1}{r^2} \right). \tag{4.27}$$

Velocity and pressure vary in opposite directions in the potential vortex. This is a typical statement of the Bernoulli equation (Fig. 4.11). Near zero, both quantities grow arbitrarily in magnitude, which is a consequence of our assumptions, in particular the absence of friction. We use (4.26) and (4.27) only for $r \geq r_1$. For a viscous flow medium, friction plays the crucial role near $r = 0$. The shear stresses there would grow arbitrarily with a velocity distribution corresponding to the potential vortex. Nature helps itself, so to speak, and the medium rotates instead like a rigid body (angular velocity $\omega =$ constant, shear stress $\tau(r) \equiv 0$):

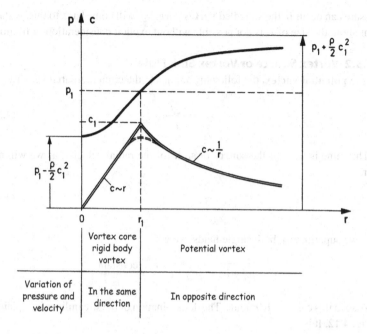

Fig. 4.11 Velocity and pressure in the vortex

$$c = \omega r = \frac{c_1}{r_1} r.$$ (4.28)

To calculate the associated pressure, we can use the equation of forces in the radial direction (4.24). Moreover, the relation (4.24) can be used whenever friction enters only through shear stresses in the tangential direction and not in the normal direction.

$$\frac{1}{\varrho} \frac{dp}{dr} = \frac{c_1^2}{r_1^2} r$$

leads to the solution ($r = r_1, p = p_1$)

$$p = p_1 + \frac{\varrho}{2} \frac{c_1^2}{r_1^2}\left(r^2 - r_1^2\right), \quad r \le r_1.$$ (4.29)

The pressure distribution (4.29) merges at $r = r_1$ with continuous tangent into (4.27). For $r < r_1$ speed and pressure vary in the same direction. Considerable low pressure can occur in the so-called **vortex core**. We will come back to this. A statement about the size of r_1 is not possible without explicit consideration of friction.

4.1.3.2 Vortex Source or Vortex Sink Flow

For the **potential vortex,** the following applies to the circumferential velocity

$$c_u = \frac{c_{u1} r_1}{r}. \tag{4.30}$$

The same is true for the **source** or **sink** for the radial velocity, as we will add later,

$$c_r = \frac{c_{r1} r_1}{r}. \tag{4.31}$$

If we superimpose both single fields, we get

$$c = \sqrt{c_r^2 + c_u^2} = \sqrt{c_{u1}^2 + c_{r1}^2}\,\frac{r_1}{r} = \frac{\text{constant}}{r}. \tag{4.32}$$

Also in this case $c \sim 1/r$ holds. The determination of the streamlines is explained in Fig. 4.12. It is

$$\tan\alpha = \frac{c_r}{c_u} = \frac{c_{r1}}{c_{u1}} = \text{constant} = \frac{dr}{r d\varphi}.$$

The integration leads to logarithmic spirals ($r = r_1, \varphi = \varphi_1$)

$$r = r_1 \exp\frac{c_{r1}}{c_{u1}}\left(\varphi - \varphi_1\right). \tag{4.33}$$

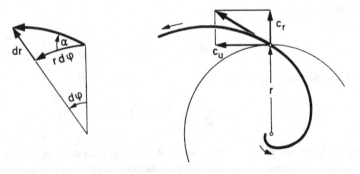

Fig. 4.12 Determination of the streamlines of the vortex source

Fig. 4.13 Case distinctions for vortex source and sink

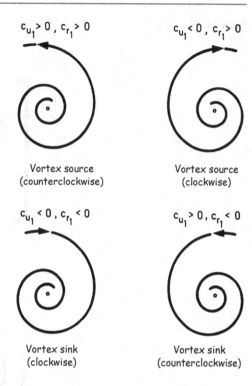

$c_{u_1} > 0 , c_{r_1} > 0$

Vortex source
(counterclockwise)

$c_{u_1} < 0 , c_{r_1} > 0$

Vortex source
(clockwise)

$c_{u_1} < 0 , c_{r_1} < 0$

Vortex sink
(clockwise)

$c_{u_1} > 0 , c_{r_1} < 0$

Vortex sink
(counterclockwise)

Figure 4.13 explains the various case distinctions as far as the direction of rotation or the source or sink property is concerned. Such vortex flows occur very frequently in nature and technology, although the magnitudes can be quite different. We remind of whirlwinds, high and low pressure areas of meteorology and spiral nebulae of astrophysics. As a simple example of application, we cite the flow in a **cyclone.** In Fig. 4.14 such a device is sketched in plan and side view. A gas flow loaded with particles (for example, air containing dust) enters a circular path tangentially at A. A vortex-sink-flow occurs, where the gas is sucked off in the dip-tube B, while the particles are thrown outwards by the centrifugal force and caught at the bottom. The radial pressure difference mentioned earlier plays a significant role in this process.

4.1.3.3 Rotational Movement Taking into Account Gravity
We are thinking here, for example, of the outflow vortex in a container with a free surface. The outflow causes a lowering of the liquid level with simultaneous rotary

Fig. 4.14 Flow in the cyclone

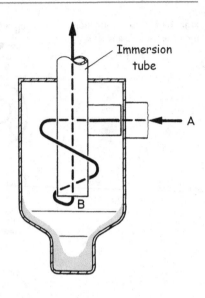

Immersion tube

A

B

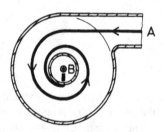

A

B

motion. In plain view, there is a good approximation of a vortex sink flow of the type discussed above. Figure 4.15 contains the terms that occur. We apply Bernoulli's equation for ϱ = constant along the streamline of $1 \rightarrow 2$:

$$\frac{c_1^2}{2} + \frac{p_1}{\varrho} + gh_1 = \frac{c_2^2}{2} + \frac{p_2}{\varrho} + gh_2. \tag{4.34}$$

At the free surface is $p_1 = p_2 = p$. If we let the point $1 \rightarrow \infty$, then $c_1 \rightarrow 0$, $h_1 \rightarrow H$. Let the index 2 be variable, neglecting the vertical velocity component becomes:

$$c_2 = c = \text{constant} \, / \, r = K \, / \, r, h_2 = z.$$

Fig. 4.15 Vortex sink flow with free surface in the gravitational field

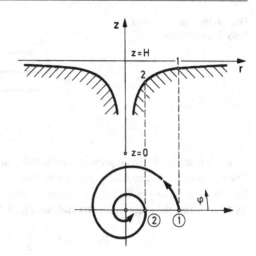

(4.34) changes to

$$gH = \frac{c^2}{2} + gz,$$

Also

$$H - z = \frac{c^2}{2g} = \frac{K^2}{2gr^2}. \tag{4.35}$$

So there is a lowering of the free surface in the funnel $\sim 1/r^2$.

4.1.3.4 Different Pressure Terms and Their Measurement

We start from the Bernoulli equation at $\varrho = $ constant in the gravitational field:

$$p + \frac{\varrho}{2}c^2 + \varrho gz = \text{constant}. \tag{4.36}$$

We designate herein in order

$$p = p_{\text{stat}} = \text{static pressure},$$

$$\frac{\varrho}{2}c^2 = p_{\text{dyn}} = \text{dynamic pressure}.$$

Fig. 4.16 Flow around a
body. Pressure term

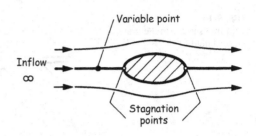

It is a coupling between pressure and velocity at each point of the velocity field. The constant is determined by suitable reference values on the respective stream-line. We discuss this in the special case of the **flow around problem** without grav-ity field. Along the **stagnation streamline** (Fig. 4.16) applies

$$p_\infty + \frac{\varrho}{2}c_\infty^2 = p + \frac{\varrho}{2}c^2 = p_0.$$

p_0 is the pressure at the stagnation point and is referred to as the resting pressure or total pressure, that is,

$$p_{stat} + p_{dyn} = p_{ges}. \tag{4.37}$$

It should be noted that if the flow is created by suction from a vessel or from the atmosphere, the resting pressure is given by the pressure in the vessel or the atmo-sphere.

The **measurement of** the **static pressure** p *is* most simply done with a **wall borehole** (Fig. 4.17) or with a **static probe** (Fig. 4.18). In the latter case, holes for taking the pressure are distributed around the circumference. These must be a suf-ficient distance from the probe tip and shaft to allow the disturbance caused by the body to decay to there. In both cases a flow boundary layer occurs. In it, the pres-sure transverse to the flow direction is practically constant, it is imposed on the boundary layer from the outside! Therefore, the static pressure of the **external flow** can be measured with these methods, because this is what matters.

For these measurements, the **relationship** between **pressure** and **head** in the pressure gauge is important. Figure 4.19 shows the incoming designations. It is

$$p' = p + g\varrho_1 h' = p_1 + g\varrho_2 h,$$

$$p - p_1 = g\varrho_2 h - g\varrho_1 h'$$

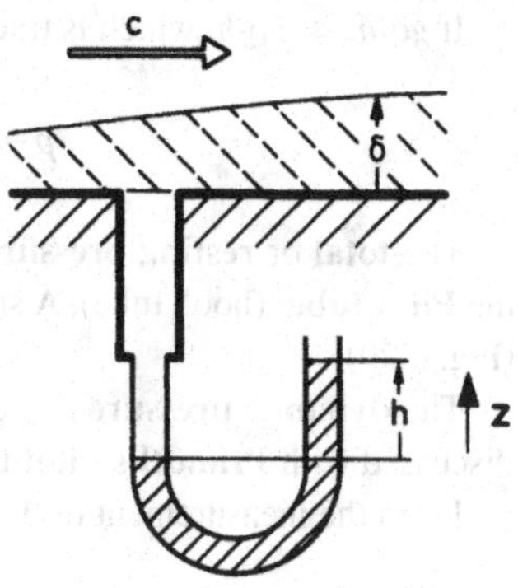

Fig. 4.17 Wall drilling (static pressure)

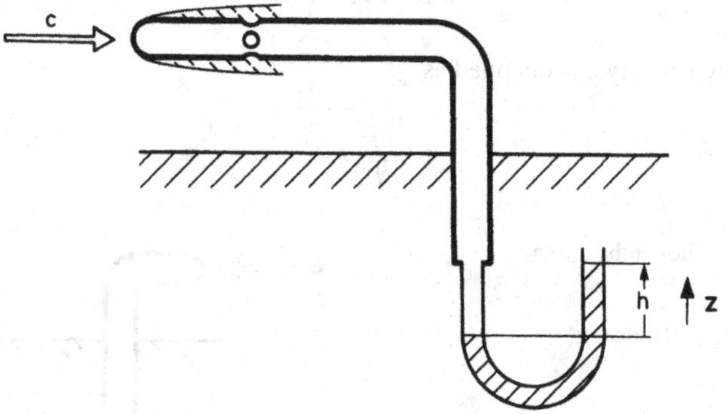

Fig. 4.18 Static probe (static pressure)

Fig. 4.19 Relationship between pressure and head in the pressure gauge

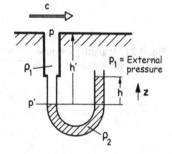

If $g\varrho_1 h' \ll g\varrho_2 h$, which is true in most cases

$$p - p_1 = \Delta p = g\varrho_2 h = g\varrho h. \tag{4.38}$$

The **total** or **resting pressure** p_0 can be measured by damming up the flow in the **Pitot tube**[2] (hook tube). A stagnation point is created in the inlet cross-section (Fig. 4.20).

The **dynamic pressure** p_{dyn} can be determined by combining the two methods discussed with **Prandtl's pitot tube** (Fig. 4.21).

From the measurement of the difference

$$P_{ges} - P_{stat} = P_{dyn}$$

the flow velocity c is obtained as

$$c = \sqrt{\frac{2}{\varrho} P_{dyn}}. \tag{4.39}$$

Fig. 4.20 Pitot tube (total pressure)

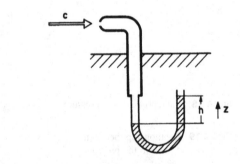

Fig. 4.21 Prandtl pitot tube (dynamic pressure)

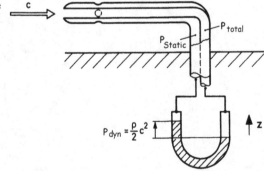

---

[2] H. Pitot, 1695–1771.

ϱ is the density of the flowing medium. With Prandtl's pitot tube one can determine the flow velocity. It should be noted that there is a **non-linear** relationship between p_{dyn} and c (4.39). If the flowing medium is **air** ($\varrho = 1.226\,kg/m^3$), the following applies

$$c = 1.28\sqrt{p_{dyn}}\,\frac{m}{s}, \quad p_{dyn}\ \text{in}\ \frac{N}{m^2} = Pa, \tag{4.40}$$

that is, $1\dfrac{N}{m^2} = 10^{-5}\,bar$ corresponds $c = 1.28\dfrac{m}{s}$,

$1cmWS \approx 100\dfrac{N}{m^2} = 10^{-3}\,bar$ on the other hand $c = 12.8\dfrac{m}{s}$.

If the flowing medium is water ($\varrho = 10^3\,kg/m^3$), then

$$c = 0.045\sqrt{p_{dyn}}\,\frac{m}{s}, \quad p_{dyn}\ \text{in}\ \frac{N}{m^2} = Pa,$$

that is, $1\dfrac{N}{m^2} = 10^{-5}\,bar$ corresponds $c = 4.5\dfrac{cm}{s}$,

$100\dfrac{N}{m^2} = 10^{-3}\,bar$ on the other hand $c = 45\dfrac{cm}{s}$. \quad (4.41)

4.1.3.5 Discharge from a Container

We first treat the **incompressible** case and trace a flow filament from the fluid surface (1) to the outlet (2) (Fig. 4.22). The Bernoulli equation is

$$\frac{c_1^2}{2} + \frac{p_1}{\varrho} + gz_1 = \frac{c_2^2}{2} + \frac{p_2}{\varrho} + gz_2. \tag{4.42}$$

If the cross-section 1 is much larger than the cross-section 2, the continuity provides

Fig. 4.22 Outflow of an incompressible medium from a container

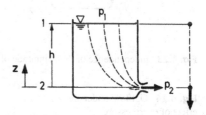

$$\frac{c_1}{c_2} = \frac{A_2}{A_1} \ll 1,$$

and we can delete $c_1^2 / 2$ in (4.42). In this case we speak of a **large** reservoir. In (1), a continuous inflow is required to keep the level constant. For the outflow velocity c_2 comes

$$c_2 = \sqrt{\frac{2}{\varrho}(p_1 - p_2) + 2gh}. \tag{4.43}$$

We consider two **special cases**. If $p_1 = p_2$ is $c_2 = \sqrt{2gh}$. This is the Torricellian formula. The same velocity occurs as in free fall from the height h and at the initial velocity $c_1 = 0$ because of the absence of friction (Fig. 4.22). It is further noteworthy that c_2 is independent of the direction of outflow. Figure 4.23 explains this by following one flow filament.

The second special case is the outflow under the effect of an overpressure, that is, without the influence of gravity (Fig. 4.24). The pressure energy is converted into kinetic energy and thus into velocity.

$$c_2 = \sqrt{\frac{2}{\varrho}(p_1 - p_2)} = \sqrt{\frac{2\Delta p}{\varrho}}.$$

We apply this relation to atmospheric motions. If we take as pressure level $\Delta p = 10$ mbar $= 10^3$ Pa, then with $\varrho = 1.226\,\text{kg/m}^3$

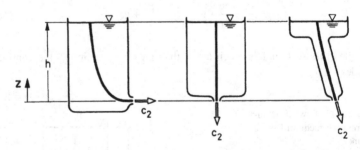

Fig. 4.23 Independence of the magnitude of the velocity from the discharge direction

Fig. 4.24 Outflow under the effect of overpressure

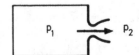

$$c_2 \approx 40 \frac{m}{s} \approx 145 \frac{km}{h}.$$

This is a considerable speed for the relatively small pressure difference. For larger pressure differences, we must take **compressibility** into account. We then speak of gas dynamics. The Bernoulli equation (4.12a) gives for horizontal motion and with $c_1 = 0$ in the reservoir (Fig. 4.24)

$$c_2 = \sqrt{2 \int_{p_2}^{p_1} \frac{dp}{\varrho}}. \tag{4.44}$$

The determination of the outflow velocity is thus based on the calculation of the integral that already occurred earlier

$$\int_{p_2}^{p_1} \frac{dp}{\varrho}.$$

If we also assume entropy here, then from (4.44) with (4.23d) we get

$$c_2 = \sqrt{2 \frac{\kappa}{\kappa-1} \frac{p_1}{\varrho_1} \left[1 - \left(\frac{p_2}{p_1} \right)^{\frac{\kappa-1}{\kappa}} \right]} = \sqrt{2 \frac{\kappa}{\kappa-1} \frac{R}{m} T_1 [\ldots]} = \sqrt{2 c_p T_1 [\ldots]}. \tag{4.45}$$

This is the formula of **Saint-Venant**[3] and **Wantzell**.[4] It represents the outflow velocity c_2 as a function of the **boiler** or **quiescent values** (p_1, ϱ_1, T_1) and the **back pressure** p_2. The **shape of** the nozzle connected to the boiler plays a major role in the realization of this velocity. This enters via the continuity equation, which has not been considered so far. We first consider (4.45). With fixed quiescent values, follows for $p_2/p_1 \to 0$ the **maximum velocity**

$$c_{2max} = \sqrt{2 \frac{\kappa}{\kappa-1} \frac{p_1}{\varrho_1}} = \sqrt{2 \frac{\kappa}{\kappa-1} \frac{R}{m} T_1}. \tag{4.46}$$

Under atmospheric conditions follows for this maximum velocity:

$$\kappa = 1.40, \quad p_1 = 1 bar, \quad \varrho_1 = 1.226 \frac{kg}{m^3},$$

$$c_{max} \approx 755 \frac{m}{s}. \tag{4.47}$$

---

[3] A. Barré de Saint-Venant, 1797–1886.
[4] P.L. Wantzell, 1814–1848.

This is a remarkable result, which clearly shows the influence of compressibility. The value (4.47) can be increased in a roundabout way via the quiescent values. Equation (4.46) offers essentially two possibilities. If one heats up the boiler, then $c_{max} \sim \sqrt{T_1}$. Here again the typical root dependence on temperature occurs. More effective, however, is the transition to lighter gases, because $c_{max} \sim 1/\sqrt{m}$. The transition from air to hydrogen provides a factor of 4 for the velocity. The extreme case $p_2/p_1 \to 0$ can be realized in two ways:

1. We hold p_1 fixed, for example, $=1$ bar, and evacuate a container, that is, $p_2 = 0$. There is then an inflow into the vacuum (Fig. 4.25)
2. We hold p_2 fixed, for example, $=1$ bar, and charge a boiler, that is, $p_1 \to \infty$. Then an outflow occurs (Fig. 4.26)

Both cases are in use to generate high speeds.

4.1.3.6 Gasdynamic Considerations. The Flow in the Laval Nozzle.[5] The Normal Shock

In order to understand the flow processes that occurred in the last section, we must deal with the concept of the speed of sound. This is another characteristic of compressible flows.

We define the **speed of sound** as the propagation velocity of small perturbations of the state variables (=sound) in a stationary compressible medium: it is a **signal velocity**, distinct different from the flow velocity. We study the wave propagation in a channel of constant cross section, a so-called **shock wave tube**. In the initial state, it is divided into two chambers by a membrane. On the right is the low pressure part and on the left the higher pressure (Fig. 4.27). If the diaphragm is removed, compression runs into the low pressure part and dilution into the high pressure part. If

Fig. 4.25 Flow into the vacuum

Fig. 4.26 Discharge from a boiler under high pressure

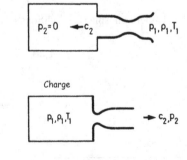

---

[5] C.G.P. de Laval, 1845–1913.

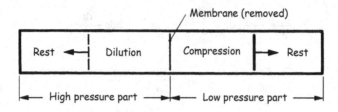

Fig. 4.27 Schematic of a shock wave tube

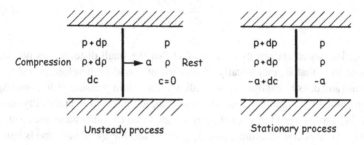

Fig. 4.28 Right-hand compression wave

the disturbances are small, the signals run at the speed of sound ($=a$). We consider the environment of the wave front traveling to the right (Fig. 4.28). This is an **unsteady** process which can be made **stationary** by the superposition of -a. We apply to this the basic equations of the flow filament theory and linearize.

Continuity with a constant cross-section:

$$-\varrho a = (\varrho + d\varrho)(-a + dc) = -\varrho a - a d\varrho + \varrho dc + \dots,$$

$$\frac{d\varrho}{\varrho} = \frac{dc}{a}. \tag{4.48}$$

Bernoulli's equation:

$$\frac{c^2}{2} + \int^p \frac{dp}{\varrho} = \text{constant},$$

$$\frac{a^2}{2} + \int^p \frac{dp}{\varrho} = \frac{(-a + dc)^2}{2} + \int^{p+dp} \frac{dp}{\varrho},$$

$$\frac{a^2}{2} = \frac{a^2}{2} - adc + \int_{p}^{p+dp} \frac{dp}{\varrho} + \ldots = \frac{a^2}{2} - adc + \frac{dp}{\varrho} + \ldots,$$

$$adc = \frac{dp}{\varrho}. \tag{4.49}$$

We combine statements (4.48) and (4.49) together:

$$a^2 = \frac{dp}{d\varrho} = \left(\frac{\partial p}{\partial \varrho}\right)_s. \tag{4.50}$$

The last statement results from the fact that the small disturbances propagate without loss, that is, entropically. The speed of sound is therefore bound to the pressure and density changes in the medium. If a certain pressure disturbance Δp is accompanied by a small change in density $\Delta \varrho$, the medium is practically incompressible and the speed of sound (4.50) is high. If, on the other hand, the change in density is considerable, compressibility prevails and the speed of sound is low.

With the isentropic change of state

$$\frac{p}{p_1} = \left(\frac{\varrho}{\varrho_1}\right)^\kappa$$

and the ideal gas equation becomes from (4.50)

$$a^2 = \left(\frac{\partial p}{\partial \varrho}\right)_s = \kappa \frac{p}{\varrho} = \kappa \frac{R}{m} T. \tag{4.51}$$

Again, the typical proportionalities result $a \sim \sqrt{T}$, $a \sim 1/\sqrt{m}$, which we have already derived for the maximum velocity. The dependence on the molar mass m is serious. It applies: $T = 300K$

Gas	O_2	N_2	H_2	Air
m in g/mol	32	28.016	2.016	29
a in m/s	330	353	1316	347

a is thus a suitable reference velocity for all compressible flows. The ratio (flow velocity)/(sound velocity) is a characteristic ratio and is named in honour of **Ernst Mach**[6] called

---

[6] E. Mach, 1838–1916.

$$\frac{c}{a} = M = Mach\ number. \tag{4.52}$$

This designation was introduced by Ackeret[7] in 1928. One distinguishes thereafter

Subsonic flows with $M < 1$ and

Supersonic flows with $M > 1$.

Figure 4.29 shows this in somewhat more detail. The following special cases occur: $M^2 \ll 1$ describes the incompressible flows, on the other hand, $M^2 \gg 1$ the so-called hypersonic flows and the $M \lessgtr 1$ near-sonic or transonic flows. This distinction has proved to be expedient.

Euler's equation gives

$$c\frac{dc}{dx} = -\frac{1}{\varrho}\frac{dp}{dx} = -\frac{1}{\varrho}\frac{dp}{d\varrho}\frac{d\varrho}{dx} = -\frac{a^2}{\varrho}\frac{d\varrho}{dx},$$

$$\frac{1}{\varrho}\frac{d\varrho}{dx} = -M^2\frac{1}{c}\frac{dc}{dx}. \tag{4.53}$$

The relative density change is proportional to the relative velocity change along the flow filament. The proportionality factor is M^2.

For $M \lessgtr 1$ the relative density change is $\lessgtr$ as the relative velocity change.

For $M \gtrless 1$ is the relative density change $\gtrless$ as the relative velocity change. For example, for $M = 10$, the proportionality factor follows 100.

In **incompressible flow** $M^2 \ll 1$, the change in velocity predominates that of the state variables p, ϱ, T. In **hypersonic** $M^2 \gg 1$ it is the other way round. In the **vicinity of sound**, all changes are of the same order of magnitude.

Fig. 4.29 Assignment of the different flows to the Mach number

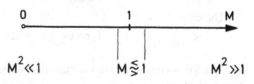

[7] J. Ackeret, 1898–1981.

Fig. 4.30 Options for the cross-section course

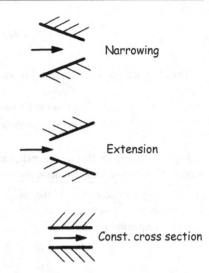

Narrowing

Extension

Const. cross section

The continuity equation shows the influence of the change in cross-section (Fig. 4.30). If we differentiate it along the flow filament, we get

$$\frac{1}{\varrho}\frac{d\varrho}{dx} + \frac{1}{c}\frac{dc}{dx} + \frac{1}{A}\frac{dA}{dx} = 0.$$

If we take into account (4.53), we get

$$\frac{1}{c}\frac{dc}{dx} = \frac{1}{M^2 - 1}\frac{1}{A}\frac{dA}{dx}. \tag{4.54}$$

Herein we take $A(x)$ as given, but $c(x)$ and $M(x)$ as unknown. (4.54) immediately provides a **qualitative** discussion of the flow in a nozzle. Its purpose is to accelerate the flow, that is, $dc/dx > 0$.

For	$M < 1$	requires this	$dA/dx < 0,$
Therefore	$M > 1$		$dA/dx > 0$
For	$M = 1$	is necessary	$dA/dx = 0$ (Fig. 3.30).

If we push these three partial results together, we inevitably arrive at the flow in the **Laval nozzle** (Fig. 4.31). In a convergent inlet the subsonic flow is accelerated; at the narrowest cross-section is the sonic passage. In the following divergent part, the supersonic flow is further accelerated. The latter is a direct consequence of Eq. (4.53). In the supersonic flow, the decrease in density predominates the increase in

Fig. 4.31 Laval nozzle

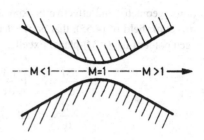

velocity. Since $\dot{m} = \varrho cA = \text{constant}$, therefore A must increase here in the direction of flow.

To **quantify** the flow, we rewrite the differential equation (4.54) to the two functions $M(x)$ and $A(x)$. We differentiate (4.52):

$$\frac{dc}{c} = \frac{da}{a} + \frac{dM}{M}$$

and use Eqs. (4.50), (4.51) and (4.53):

$$2\frac{da}{a} = \frac{dp}{p} - \frac{d\varrho}{\varrho} = \frac{a^2}{p}d\varrho - \frac{d\varrho}{\varrho} = (\kappa - 1)\frac{d\varrho}{\varrho} = -M^2(\kappa - 1)\frac{dc}{c},$$

$$\frac{dM}{M} = \left(1 + \frac{\kappa - 1}{2}M^2\right)\frac{dc}{c}.$$

If we take this into account in (4.54).

$$\frac{1}{M}\frac{dM}{dx} = \frac{1 + \frac{\kappa - 1}{2}M^2}{M^2 - 1}\frac{1}{A}\frac{dA}{dx}. \tag{4.55}$$

This is an ordinary differential equation of first order for $M = M(x)$, which can be solved by separation of variables. With the condition $M^* = 1$, $A(M^* = 1) = A^*$ becomes

$$\frac{A}{A^*} = \frac{1}{M}\left[1 + \frac{\kappa - 1}{\kappa + 1}(M^2 - 1)\right]^{\frac{\kappa + 1}{2(\kappa - 1)}}. \tag{4.56}$$

Here implicitly is given the Mach-number as a function of nozzle-cross-section, if at the narrowest point A^* the speed of sound prevails. We want to obtain an overview of all possible flows in a Laval nozzle as a function of the boundary condi-

tions (geometry and effective backpressure). For this purpose we determine the
direction field of (4.55), that is, we determine the slope of the solution curve in
each point of the (x, M)-plane. Excellent directional elements are:

$$\frac{dM}{dx} = 0, \quad \text{if } \frac{dA}{dx} = 0, while\, M \neq 1,$$

$$\frac{dM}{dx} = \infty, \quad \text{if } M = 1, while\, \frac{dA}{dx} \neq 0.$$

Singular points are located where the factor of dM/dx is zero or infinity. Only
there can intersect integral curves. $M = 0$ corresponds to $A \rightarrow \infty$, that is, the boiler.
Here infinite directions of progression are possible. $M = 1$ and $dA/dx = 0$ lead to an
indefinite expression in (4.55). The application of Bernoulli-L'Hospital's[8] rule
gives

$$\left(\frac{dM}{dx}\right)_{1,2} = \pm\sqrt{\frac{\kappa+1}{4}\frac{\dfrac{d^2A}{dx^2}}{A^*}}. \tag{4.57}$$

There are **two** directions of progression if $d^2A/dx^2 > 0$, that is, at the narrowest
cross section. It is there a **saddle point**. If, on the other hand, $d^2A/dx^2 < 0$ which
corresponds to the maximum of the function $A(x)$, there is no real direction of progression. There is a **vortex point**.

After these preparations, the field of integral curves can be drawn immediately
(Fig. 4.32). By varying the pressure at the nozzle end, the different flows can be
realized. If there is a small difference between the boiler pressure and the back
pressure (A), we obtain a subsonic nozzle. (B) corresponds to the case where at the
narrowest cross-section the speed of sound is reached but not passed. If the pressure is reduced further (C), it is immediately apparent that a **steady** flow is no
longer possible. A so-called (normal) **compression shock** occurs, in which the
state variables change discontinuously. The velocity decreases to subsonic; pressure, density and temperature increase. With further pressure reduction, the shock
moves to the nozzle end (D). Between D and E an oblique shock occurs at the
outlet. E is borderline case of ideal Laval-nozzle. Here exists a parallel jet at outlet.
If the pressure is reduced further (F), an expansion occurs there (Fig. 4.33). This
heuristic description shows the variety of possible flow processes depending on the
boundary conditions. For the moment, we only follow steady flows. We will return
to the calculation of the compression shocks at the end of this section.

---

[8] G.Fr.A. de L'Hospital, 1661–1704.

Fig. 4.32 Mach number curve in the Laval nozzle at different back pressures

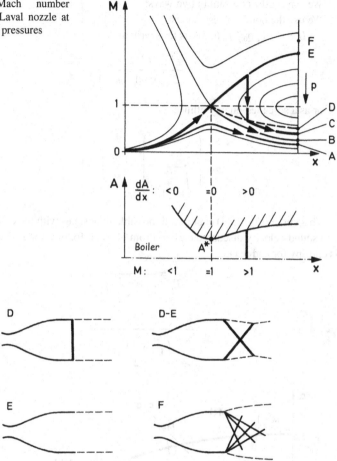

Fig. 4.33 Influence of back pressure on the flow pattern in the Laval nozzle

We thus have the Mach number $M(x)$ at every point in the nozzle. The calculation of $p(x)$, $\varrho(x)$ and $T(x)$ is done with the **Bernoulli equation**. For **isentropic** changes of state

$$\frac{c^2}{2}+\frac{\kappa}{\kappa-1}\frac{p}{\varrho}=\frac{c^2}{2}+\frac{a^2}{\kappa-1}=\frac{c^2}{2}+c_pT=\text{constant.} \tag{4.58}$$

We can set the constant in two ways:
1. We use the boiler or rest values[9]

   $c = 0$, a_0 , p_0 , ϱ_0 , T_0 (4.58) is thus written

$$\frac{c^2}{2} + \frac{a^2}{\kappa - 1} = \frac{a_0^2}{\kappa - 1} \tag{4.59}$$

or as a so-called energy ellipse

$$\left(\frac{c}{\sqrt{\frac{2}{\kappa - 1}}a_0}\right)^2 + \left(\frac{a}{a_0}\right)^2 = 1. \tag{4.60}$$

This curve (Fig. 4.34) captures all possible flow states with the changes in flow and sound velocity discussed earlier. From (4.59) it follows for $T(x)$ then with the is entropy for and $p(x)$:

$$\frac{T}{T_0} = \frac{1}{1 + \frac{\kappa - 1}{2} M^2}, \tag{4.61a}$$

$$\frac{\varrho}{\varrho_0} = \frac{1}{\left(1 + \frac{\kappa - 1}{2} M^2\right)^{\frac{1}{\kappa - 1}}}, \tag{4.61b}$$

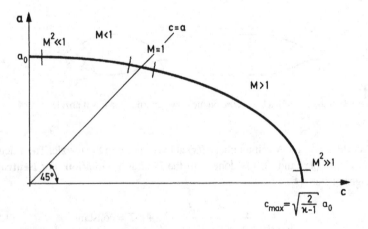

Fig. 4.34 Energy ellipse in the (c,a)-plane

[9] In the following we denote the rest values by the index 0, as it is common in gas dynamics.

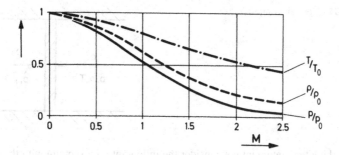

Fig. 4.35 T, ϱ, p as functions of the Mach number

$$\frac{p}{p_0} = \frac{1}{\left(1 + \dfrac{\kappa - 1}{2} M^2\right)^{\frac{\kappa}{\kappa - 1}}}. \tag{4.61c}$$

Figure 4.35 explains the different variation of the state variables. The critical values at $M = 1$ are characteristic.

$$\frac{T^*}{T_0} = \frac{2}{\kappa + 1} = 0.833, \quad \frac{\varrho^*}{\varrho_0} = \left(\frac{2}{\kappa + 1}\right)^{\frac{1}{\kappa - 1}} = 0.634,$$

$$\frac{p^*}{p_0} = \left(\frac{2}{\kappa + 1}\right)^{\frac{\kappa}{\kappa - 1}} = 0.528. \tag{4.62}$$

The figures refer to $\kappa = 1.40$.

2. We can use the **critical values** as reference **values:** $c = a = a^*, p^*, \varrho^*, T^*$. Besides $M = c/a$, the **critical Mach number** $M^* = c/a^*$ occurs. For invariant rest values is a^* constant. The normalization with the critical sound velocity has thus the practical advantage that in the denominator of the Mach number no local, that is, variable, sound velocity stands any more.

We have the following range of variability:

M	0	1	∞
M^*	0	1	$\sqrt{\dfrac{\kappa + 1}{\kappa - 1}}$

Fig. 4.36 Normal compression shock

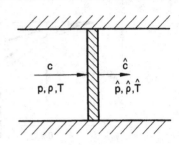

For the latter statement we consider the inflow into the vacuum with the maximum velocity (4.46)

$$M_{max}^* = \frac{c_{max}}{a^*} = \frac{\sqrt{\frac{2}{\kappa-1}} a_0}{a^*} = \sqrt{\frac{2}{\kappa-1}} \sqrt{\frac{T_0}{T^*}} = \sqrt{\frac{\kappa+1}{\kappa-1}} \; (= 2.45 \text{ for air}).$$

This results in the following calculation procedure: Given the known nozzle geometry $A(x)$, we determine $M(x)$. Subsequently (4.61a, 4.61b, 4.61c) gives $T(x)$, $\varrho(x)$ and $p(x)$.

We now come to the treatment of the **normal compression shock**. We consider a one-dimensional steady flow in the flow filament of constant cross section (Fig. 4.36). We apply the conservation laws for mass, momentum[10] and energy across the shock front.

$$\text{Continuity}: \quad \varrho c = \hat{\varrho}\hat{c},$$

$$\text{Momentum}: \quad p + \varrho c^2 = \hat{p} + \hat{\varrho}\hat{c}^2,$$

$$\text{Energy}: \quad h + \frac{1}{2}c^2 = \hat{h} + \frac{1}{2}\hat{c}^2,$$

with the enthalpy $h = c_p T = \dfrac{\kappa}{\kappa-1}\dfrac{p}{\varrho}$ this becomes

$$\frac{\kappa}{\kappa-1}\frac{p}{\varrho} + \frac{1}{2}c^2 = \frac{\kappa}{\kappa-1}\frac{\hat{p}}{\hat{\varrho}} + \frac{1}{2}\hat{c}^2.$$

This system has two solutions with known initial values (c, p, ϱ, T):

---

[10] This is done in anticipation of Sect. 4.3.1.

$$\frac{\hat{c}}{c} = \frac{\varrho}{\hat{\varrho}} = \begin{cases} 1, \\ 1 - \dfrac{2}{\kappa+1}\left(1 - \dfrac{\kappa p}{\varrho c^2}\right), \end{cases}$$

$$\frac{\hat{p}}{p} = \begin{cases} 1, \\ 1 + 2\dfrac{\kappa}{\kappa+1}\left(\dfrac{c^2\varrho}{\kappa p} - 1\right). \end{cases}$$

The first solution is the identity, the second gives the changes of the state over the shock. With the speed of sound a (4.51), the Mach number M (4.52) and the entropy s of the mass unit, we get

$$\frac{\hat{c}}{c} = \frac{\varrho}{\hat{\varrho}} = 1 - \frac{2}{\kappa+1}\left(1 - \frac{1}{M^2}\right),$$

$$\frac{\hat{p}}{p} = 1 + 2\frac{\kappa}{\kappa+1}\left(M^2 - 1\right),$$

$$\frac{\hat{T}}{T} = \frac{\hat{p}}{p}\frac{\varrho}{\hat{\varrho}},$$

$$\frac{\hat{s} - s}{c_v} = \ln\left\{\frac{\hat{p}}{p}\left(\frac{\varrho}{\hat{\varrho}}\right)^\kappa\right\} = \ln\left\{\left[1 + 2\frac{\kappa}{\kappa+1}\left(M^2 - 1\right)\right]\left[1 - \frac{2}{\kappa+1}\left(1 - \frac{1}{M^2}\right)\right]^\kappa\right\}.$$

Figures 4.37 and 4.38 show the change in the state variables in the case of a normal shock. Since an impact $\hat{s} - s > 0$ must be (Fig. 4.38), a normal shock can only occur in supersonic flow $M > 1$. This is a compression (Fig. 4.37) with transition from supersonic to subsonic. This follows from the Prandtl relation for the critical Mach numbers:

$$M^* \cdot \hat{M}^* = 1.$$

It is an elementary consequence of the derived shock equations. Characteristic is the behavior of the rest values. If we think of the medium as being at rest before and after the shock, the energy theorem over the shock is as follows

Fig. 4.37 The shock quantities for a normal shock as a
function of M ($\kappa = 1.40$)

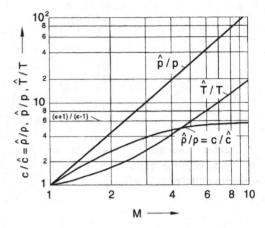

Fig. 4.38 The entropy in the
normal compression shock as
a function of M ($\kappa = 1.40$)

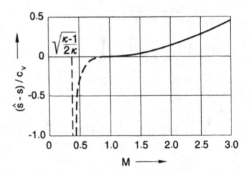

$$c_p T_0 = c_p T + \frac{1}{2} c^2 = c_p \hat{T} + \frac{1}{2} \hat{c}^2 = c_p \hat{T}_0,$$

that is,

$$T_0 = \hat{T}_0, \quad a_0 = \hat{a}_0, \quad a^* = \hat{a}^*.$$

For pressure and density, isentropic deceleration is performed before and after
the impact. If one uses an isothermal comparison process, then because of

$$\hat{s}_0 - s_0 = \hat{s} - s = -\left(c_p - c_v\right)\ln\frac{\hat{p}_0}{p_0} = -\left(c_p - c_v\right)\ln\frac{\hat{\varrho}_0}{\varrho_0},$$

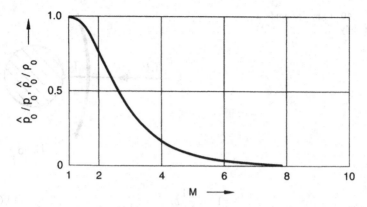

Fig. 4.39 Decrease in resting pressure and resting density for normal shock as a function of M ($\kappa = 1.40$)

$$\frac{\hat{p}_0}{p_0} = \frac{\hat{\varrho}_0}{\varrho_0} = \left[1 + 2\frac{\kappa}{\kappa+1}\left(M^2 - 1\right)\right]^{-\frac{1}{\kappa-1}}\left[1 - \frac{2}{\kappa+1}\left(1 - \frac{1}{M^2}\right)\right]^{-\frac{\kappa}{\kappa-1}}.$$

The decrease of the resting pressure is small in the vicinity of sound, but it is considerable for strong shocks at high Mach numbers (Fig. 4.39).

We hold: About the normal shock.

are increasing:	p, ϱ, T, s
are decreasing:	$c, M, p_0, \varrho_0, p^*, \varrho^*$
remain constant:	T_0, a_0, T^*, a^*

To explain the introduced terms, we give some **examples of gas dynamics**.

1. **Temperature increase at the stagnation point of** a missile. On the central streamline (Fig. 4.40), a stagnation occurs, which can lead to a significant temperature increase. We can apply (4.61a) to find the temperature at the stagnation point ($=T_0$). For $M > 1$, there is a head wave between the inflow and the body. (4.61a) holds across this shock, since this equation is identical to the energy theorem that applies to shocks:

Fig. 4.40 Flow around a blunt body

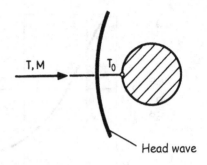

Head wave

$$\frac{T_0}{T} = 1 + \frac{\kappa - 1}{2} M^2. \tag{4.63}$$

For air, this means

$$M = 2, \frac{T_0}{T} = 1.8, \text{that is, at } T = 300\text{K} : T_0 = 540\text{K} = 267°\text{C},$$

$$M = 5, \frac{T_0}{T} = 6, T_0 = 1800\text{K} = 1527°\text{C}.$$

At the last temperatures one is already at the limit of the range of validity of the ideal gases of constant specific heats. With further increase dissociation, ionization etc. occur. These effects require energy and lead to the fact that the stagnation point temperature resulting from (4.63) is actually considerably undershoot.

2. Up to what Mach number (velocity) can a **flow** be considered **incompressible**? We require that in such a case the relative density change should be less than 1%. (4.61b) gives

$$\frac{\varrho}{\varrho_0} = \frac{1}{\left(1 + \dfrac{\kappa - 1}{2} M^2\right)^{\frac{1}{\kappa-1}}} = \frac{1}{1 + \dfrac{M^2}{2} + \ldots} = 1 - \frac{M^2}{2} + \ldots,$$

$$\left| \frac{\varrho - \varrho_0}{\varrho_0} \right| = \frac{M^2}{2} + \ldots \leq 0.01, \quad M \leq 0.14.$$

This Mach number, in the air at room temperature, leads to $c \leq 50$ m/s.

3. Determination of the **amount of leakage from a boiler** at **supercritical condition** (Fig. 4.41). The leakage forms a critical cross section A^*. We trace the mass flow to the boiler values ($\kappa = 1.40$):

Fig. 4.41 Outflow from a
boiler in the supercritical case

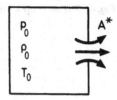

$$\dot{m} = \varrho^* c^* A^* = \varrho^* a^* A^* = \left(\frac{2}{\kappa+1}\right)^{\frac{1}{\kappa-1}} \varrho_0 \sqrt{\frac{2}{\kappa+1}} a_0 A^* = 0.58 \varrho_0 a_0 A^*;$$

with $a_0 = 347$m/s it will be

$$\frac{\dot{m}}{\varrho_0 A^*} = 0.58 \cdot 347 \frac{\text{m}}{\text{s}} = 2.10^{-2} \frac{\text{m}^3}{\text{scm}^2} = 20 \frac{\ell}{\text{scm}^2}.$$

This means that $20\,\ell$ of air of the boiler state flow through one square centimetre per second. In 10 s thus 20m^3 at 100cm^2, if critical pressure difference is maintained. These numbers illustrate large mass-through flow within the narrowest cross-section of Laval-nozzle. They give an idea of the capacity of a reservoir connected to such a nozzle.

4. **Filling a boiler**—principle of a supersonic wind tunnel. We connect a Laval nozzle to an evacuated boiler of volume V (Fig. 4.42). Let the initial state in the boiler be: p_a, ϱ_a, T_a, outside, for example, that of the atmosphere $p0$, ϱ_0, $T0$. If we remove the separating membrane, in Fig. 4.42 to be imagined for example, at the nozzle end, a transient starting process occurs similar to that in the shock wave tube. If p_a/p_0 is sufficiently small, a Laval nozzle flow is present. After a short start-up phase, the stationary flow associated with the pressure ratio p_a/p_0 is then established for a few seconds. This is the measuring time of the channel. During this phase, models, for example, airfoils in supersonic flow, can be examined in the measuring section. The flow is made visible with suitable methods and observed through the duct windows. However, in this case the vessel is gradually filled up. The state values $p(t)$, $\varrho(t)$, $T(t)$ can be calculated from the volume V and the nozzle geometry. As time progresses, that is, as $p(t)$ increases, the flow states discussed earlier passed through: expansion, oblique shock, normal shock. This normal shock moves to the nozzle throat with weakening. Thus the critical state collapses there and we have a subsonic flow until complete pressure equalization is established.

Fig. 4.42 Filling up a boiler

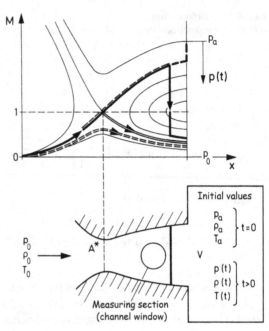

5. The **Pitot tube in supersonic flow** (Fig. 4.43) is used to measure $\hat{p}_0$. When M is known, p_0 can be calculated by the formula for loss at rest. However, if p or $\hat{p}$ **and** $\hat{p}_0$ are measured, M can be determined. For this purpose, the isentropic relation (4.61c) is used. We give the corresponding formulae, since they are important for the supersonic measurement technique:

$$\frac{\hat{p}_0}{p} = \frac{\left(\dfrac{\kappa+1}{2}M^2\right)^{\frac{\kappa}{\kappa-1}}}{\left[1+\dfrac{2\kappa}{\kappa+1}\left(M^2-1\right)\right]^{\frac{\kappa}{\kappa-1}}},$$

$$\frac{\hat{p}_0}{\hat{p}} = \frac{\left(\dfrac{\kappa+1}{2}M^2\right)^{\frac{\kappa}{\kappa-1}}}{\left[1+\dfrac{2\kappa}{\kappa+1}\left(M^2-1\right)\right]^{\frac{\kappa}{\kappa-1}}}$$

Fig. 4.43 Pitot tube in supersonic flow

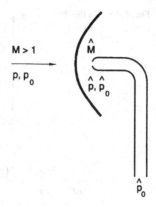

4.2   Frictionless, Plane and Spatial Flows

In the following we extend the one-dimensional flow filament theory to several independent variables using Euler's method.

4.2.1   Continuity (=Conservation of Mass)

We consider a spatially fixed control region, a cuboid with edge lengths dx, dy, dz (Fig. 4.44). A change in mass flux through the boundary leads to a change in mass inside. The mass flow through the surface in the *x-direction* is

$$d\dot{m}_x = \varrho u dy dz - \left(\varrho u + \frac{\partial(\varrho u)}{\partial x} \right) dy dz = -\frac{\partial(\varrho u)}{\partial x} dx dy dz.$$

For all three axis directions a total of

$$d\dot{m} = -\left(\frac{\partial(\varrho u)}{\partial x} + \frac{\partial(\varrho v)}{\partial y} + \frac{\partial(\varrho w)}{\partial z} \right) dx dy dz = \frac{\partial \varrho}{\partial t} dx dy dz. \qquad (4.64)$$

This equation expresses that the resulting mass flow through the surface must be reflected in a local temporal increase or decrease in mass inside. In other words, the mass can only increase in the interior by, for example, more flowing in than flowing out. (4.64) can be written in different forms:

Fig. 4.44 On the derivation of continuity

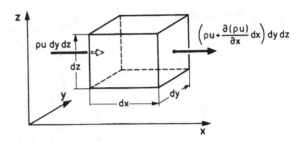

$$0 = \frac{\partial \varrho}{\partial t} + \frac{\partial (\varrho u)}{\partial x} + \frac{\partial (\varrho v)}{\partial y} + \frac{\partial (\varrho w)}{\partial z} = \frac{\partial \varrho}{\partial t} + div(\varrho w) = \frac{d\varrho}{dt} + \varrho div w. \quad (4.65)$$

From this follows the physical meaning of **divergence**

$$div w = -\frac{1}{\varrho}\frac{d\varrho}{dt}$$

as the **relative yield of** the flow field.

4.2.2   Euler's Equations of Motion

We apply Newton's fundamental law to the spatially fixed mass element (Fig. 4.45) and obtain in turn

Mass = dm = $\varrho dx dy dz$,

Acceleration = $\dfrac{dw}{dt}$,

Applied forces = Mass   and surface forces = $f dm - grad\ p dx dy dz$,

with $f = (fx, fy, fz)$ as the force related to the mass. So

$$\frac{dw}{dt} = -\frac{1}{\varrho} grad\, p + f. \quad (4.66)$$

Altogether we get the statement: Continuity and Euler's equations provide four conditions for the five unknowns =(u, v, w), p and ϱ. So, again, an additional equation (energy theorem) is needed to determine all the unknowns.

Fig. 4.45 Space fixed mass element

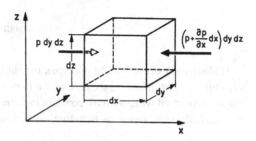

4.2.3 Plane, Stationary, Incompressible Potential Flow

This is a special case important for fluid mechanics, which will occupy us in detail. The condition ϱ = constant replaces the missing equation. If we still eliminate gravity, then:

Continuity:

$$\frac{\partial u}{\partial x} + \frac{\partial v}{\partial y} = 0, \tag{4.67}$$

Euler's equations:

$$u\frac{\partial u}{\partial x} + v\frac{\partial u}{\partial y} = -\frac{1}{\varrho}\frac{\partial p}{\partial x}, \tag{4.68a}$$

$$u\frac{\partial v}{\partial x} + v\frac{\partial v}{\partial y} = -\frac{1}{\varrho}\frac{\partial p}{\partial y}. \tag{4.68b}$$

These equations for u, v and p can be reduced to two relations for u and v. We differentiate (4.68a) by y and (4.68b) by x and subtract. If we use (4.67) for simplification, we get

$$0 = u\frac{\partial}{\partial x}\left(\frac{\partial v}{\partial x} - \frac{\partial u}{\partial y}\right) + v\frac{\partial}{\partial y}\left(\frac{\partial v}{\partial x} - \frac{\partial u}{\partial y}\right) = \frac{\mathrm{d}}{\mathrm{d}t}\left(\frac{\partial v}{\partial x} - \frac{\partial u}{\partial y}\right), \tag{4.69}$$

where the last equation takes into account that we are studying steady-state flows.

(4.69) shows that

$$\frac{\partial v}{\partial x} - \frac{\partial u}{\partial y} = \text{constant} \tag{4.70}$$

is along each streamline. In principle, the value of this constant can vary from streamline to streamline. In the case of the **flow problems around bodies** we are mainly concerned with, we have constant incident flow at infinity, that is, $u \to u_\infty$, $v \to v_\infty$. In the limit $x \to -\infty$, therefore, on each streamline from (4.70)

$$\frac{\partial v}{\partial x} - \frac{\partial u}{\partial y} = \left(\frac{\partial v}{\partial x} - \frac{\partial u}{\partial y} \right)_{x \to -\infty} = 0.$$

So the constant is zero in our case. This gives the basic equations

$$\text{Continuity}: \quad \frac{\partial u}{\partial x} + \frac{\partial v}{\partial y} = 0, \tag{4.71a}$$

$$\text{Freedom of rotation}: \quad \frac{\partial v}{\partial x} - \frac{\partial u}{\partial y} = 0. \tag{4.71b}$$

We explain the concept of **rotation** with two simple examples.

1. In the case of the **rigid body vortex** (Fig. 4.46)

$$c = \omega r, \quad u = -\omega r \sin\varphi = -\omega y, \quad v = \omega r \cos\varphi = \omega x.$$

Fig. 4.46 Rigid body vortex

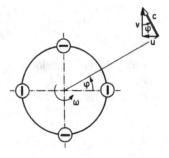

Fig. 4.47 Potential vortex

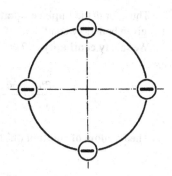

So

$$\frac{\partial v}{\partial x} - \frac{\partial u}{\partial y} = 2\omega,$$

and thus $\partial v/\partial x - \partial u/\partial y$ can be taken as a measure of the local rotation of the particle.

2. With the **potential vortex** (Fig. 4.47) is

$$c = \frac{k}{r}, \quad u = -k\frac{\sin\varphi}{r} = -k\frac{y}{x^2 + y^2}, \quad v = k\frac{\cos\varphi}{r} = k\frac{x}{x^2 + y^2},$$

and thus

$$\frac{\partial v}{\partial x} - \frac{\partial u}{\partial y} = 0,$$

that is, here is a **rotation-free** movement. By integrating the two differential equations (4.71a, 4.71b) for u and v we determine the velocity field. Then the pressure is determined with Bernoulli's equation. There are two possible solutions here.

1. We satisfy the rotational freedom (4.71b) by a potential function $\Phi(x, y)$, that is,

$$u = \frac{\partial\Phi}{\partial x}, \quad v = \frac{\partial\Phi}{\partial y}. \tag{4.72}$$

Then the **continuity** (4.71a) gives the condition

$$\frac{\partial u}{\partial x} + \frac{\partial v}{\partial y} = \frac{\partial^2\Phi}{\partial x^2} + \frac{\partial^2\Phi}{\partial y^2} = \Delta\Phi = 0. \tag{4.73}$$

Thus, for Φ the **Laplace equation**[11] (= potential equation) is to be solved under the given boundary conditions of the flow problem.

2. We satisfy **continuity** (4.71a) by a **stream function** $\Psi(x, y)$, that is,

$$u = \frac{\partial \Psi}{\partial y} = \frac{\partial \Phi}{\partial x}, \quad v = -\frac{\partial \Psi}{\partial x} = \frac{\partial \Phi}{\partial y}. \tag{4.74}$$

The **freedom of rotation** (4.71b) requires

$$\frac{\partial^2 \Psi}{\partial y^2} + \frac{\partial^2 \Psi}{\partial x^2} = \Delta \Psi = 0,$$

that is, **Laplace's equation** also applies to Ψ.

Φ and Ψ have an important physical meaning:

1. For the contour lines of the Ψ-surface, that is, the curves Ψ = constant, applies

$$\mathrm{d}\Psi = \frac{\partial \Psi}{\partial x}\mathrm{d}x + \frac{\partial \Psi}{\partial y}\mathrm{d}y = -v\mathrm{d}x + u\mathrm{d}y = 0,$$

$$\left(\frac{\mathrm{d}y}{\mathrm{d}x}\right)_{\psi=\text{constant}} = \frac{v}{u}.$$

Consequently, the curves Ψ = constant are **streamlines**.

2. For the curves Φ = constant comes accordingly

$$\mathrm{d}\Phi = \frac{\partial \Phi}{\partial x}\mathrm{d}x + \frac{\partial \Phi}{\partial y}\mathrm{d}y = u\mathrm{d}x + v\mathrm{d}y = 0,$$

$$\left(\frac{\mathrm{d}y}{\mathrm{d}x}\right)_{\Phi=\text{constant}} = -\frac{u}{v}.$$

The curves Φ = constant, the potential lines, are orthogonal to the streamlines. Φ = constant and Ψ = constant form an orthogonal network (Fig. 4.48). For the volumetric flow, in terms of the unit of width or depth, between two streamlines, the following holds (Fig. 4.49)

---

[11] P.S. Laplace, 1749–1827.

Fig. 4.48 Orthogonal network of potential-lines and streamlines

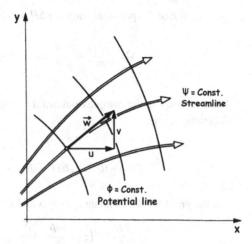

Fig. 4.49 Calculation of the volume flow between two streamlines

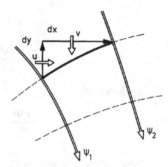

$$\dot{V}_{12} = \int_1^2 d\dot{V} = \int_1^2 (u\,dy - v\,dx) = \int_1^2 \left(\frac{\partial \Psi}{\partial y} dy + \frac{\partial \Psi}{\partial x} dx \right) = \int_1^2 d\Psi = \Psi_2 - \Psi_1. \quad (4.75)$$

That is, the difference between the Ψ–values of two streamlines in the plane case provides the volume flow per unit depth perpendicular to the image plane between them.

We now discuss general **methods of solving** the equations $\Delta\Phi = 0$ and $\Delta\Psi = 0$.

1. Any differentiable complex function

$$F(z) = F(x + iy) = H(x,y), \quad i = \sqrt{-1}$$

is a solution of the potential equation $\Delta H = 0$, because

$$\Delta H = \frac{\partial^2 H}{\partial x^2} + \frac{\partial^2 H}{\partial y^2} = F''(z) - F''(z) = 0.$$

2. If we decompose the **complex potential** $F(z)$ into real and imaginary parts, the following applies

$$F(z) = \operatorname{Re} F + \mathrm{i} \operatorname{Im} F = \Phi(x,y) + \mathrm{i} \Psi(x,y).$$

For the proof we differentiate here by x and y.

$$\frac{\partial}{\partial x} : F'(z) = \frac{\partial \Phi}{\partial x} + \mathrm{i} \frac{\partial \Psi}{\partial x},$$

$$\frac{\partial}{\partial y} : \mathrm{i} F'(z) = \frac{\partial \Phi}{\partial y} + \mathrm{i} \frac{\partial \Psi}{\partial y}.$$

that is $\dfrac{\partial \Phi}{\partial x} + \mathrm{i} \dfrac{\partial \Psi}{\partial x} = -\mathrm{i} \dfrac{\partial \Phi}{\partial y} + \dfrac{\partial \Psi}{\partial y}$,

and therefore

$$\frac{\partial \Phi}{\partial x} = \frac{\partial \Psi}{\partial y}, \quad \frac{\partial \Phi}{\partial y} = -\frac{\partial \Psi}{\partial x}.$$

These are the relations (4.72) and (4.74) which apply to potential and current functions, respectively, thus proving the assertion. They are called **Cauchy-Riemann differential equations**[12] and play an important role in the theory of functions.

The above statements 1 and 2 are of fundamental interest for fluid mechanics. Thus, if one decomposes a complex function into real and imaginary parts, one obtains the potential and flow function of a potential flow. The difficulty lies in determining those functions which satisfy the given **boundary conditions of** the flow problem.

Another property of differential equations $\Delta \Phi = 0$ and $\Delta \Psi = 0$ is their **linearity**. With Φ_1 and Φ_2 is $c_1 \Phi_1 + c_2 \Phi_2$ ($c_1, c_2 =$ constant) also a solution, because

$$\Delta(c_1 \Phi_1 + c_2 \Phi_2) = \Delta(c_1 \Phi_1) + \Delta(c_2 \Phi_2) = c_1 \Delta \Phi_1 + c_2 \Delta \Phi_2 = 0.$$

---

[12] A.L. Cauchy, 1789–1857; B. Riemann, 1826–1866.

This superposition can also be done graphically. We explain this with the example of the superposition of parallel flow (Ψ_1) and source (Ψ_2). Figure 4.50 shows the emergence of the new field with the stream function $\Psi = \Psi_1 + \Psi_2$. **Each** streamline can be taken as a material obstacle and considered as a body with the flow around it. If we take the stagnation streamline, we have a model for the flow around a blunt half-body. We will encounter this flow again in the analytical treatment.

Bernoulli's equation also holds here, as can be easily verified with the Euler equations (4.68a, b) and the freedom of rotation (4.71b):

$$p + \frac{\varrho}{2}\left(u^2 + v^2\right) = p_\infty + \frac{\varrho}{2}c_\infty^2 = p_0. \tag{4.76}$$

For the dimensionless representation we use the **pressure coefficient**

$$c_p = \frac{p - p_\infty}{\frac{\varrho}{2}c_\infty^2} = \frac{\Delta p}{q} = 1 - \left(\frac{c}{c_\infty}\right)^2. \tag{4.77}$$

Special values are $c_{p_\infty} = 0$ in the **inflow** as well as $c_{p_0} = 1$ in the **stagnation points**.

4.2.4 Examples of Elementary and Composite Potential Flows

We discuss the examples listed in Table. 4.1. At the beginning we will be brief, later we will be more detailed. We thereby gain experience about simple flow fields, which we will need later.

1. **Parallel flow**

$$F(z) = \left(u_\infty - iv_\infty\right)z = \left(u_\infty - iv_\infty\right)\left(x + iy\right) = u_\infty x + v_\infty y + i\left(u_\infty y - v_\infty x\right),$$
$$\Phi = u_\infty x + v_\infty y, \quad \Psi = u_\infty y - v_\infty x; \quad \Phi_x = u = u_\infty, \quad \Phi_y = v = v_\infty.$$

Streamlines $\Psi = $ constant : $y = \dfrac{v_\infty}{u_\infty}x + \text{constant}.$

Table 4.1 Elementary and composite potential flows

Complex potential $F(z)$	Potential $\Phi(x,y)$	Stream function $\Psi(x,y)$	Velocity u	v	c	Streamlines $\Psi = $ constant
$(u_\infty - iv_\infty)z$ Parallel flow	$u_\infty x + v_\infty y$	$u_\infty y - v_\infty x$	u_∞	v_∞	$c_\infty = \sqrt{u_\infty^2 + v_\infty^2}$	
$\dfrac{Q}{2\pi}\ln z = \dfrac{Q}{2\pi}\ln\sqrt{x^2+y^2}$ Source $Q>0$, Sink $Q<0$	$\dfrac{Q}{2\pi}\ln r = \dfrac{Q}{2\pi}\ln\sqrt{x^2+y^2}$	$\dfrac{Q}{2\pi}\varphi = \dfrac{Q}{2\pi}\arctan\dfrac{y}{x}$	$\dfrac{Q}{2\pi}\dfrac{x}{x^2+y^2}$	$\dfrac{Q}{2\pi}\dfrac{y}{x^2+y^2}$	$\dfrac{Q}{2\pi r}$	
$\dfrac{\Gamma}{2\pi}\,\mathrm{i}\ln z$ Vortex $\Gamma \gtreqless 0$ clockwise counter clockwise	$-\dfrac{\Gamma}{2\pi}\arctan\dfrac{y}{x}$	$\dfrac{\Gamma}{2\pi}\ln\sqrt{x^2+y^2}$	$\dfrac{\Gamma}{2\pi}\dfrac{y}{x^2+y^2}$	$-\dfrac{\Gamma}{2\pi}\dfrac{x}{x^2+y^2}$	$\dfrac{\Gamma}{2\pi r}$	
$\dfrac{m}{z}$ Dipole	$\dfrac{mx}{x^2+y^2}$	$-\dfrac{my}{x^2+y^2}$	$m\dfrac{y^2-x^2}{(x^2+y^2)^2}$	$-m\dfrac{2xy}{(x^2+y^2)^2}$	$\dfrac{m}{r^2}$	
$u_\infty z + \dfrac{Q}{2\pi}\ln z$ Parallel flow + Source/ Sink	$u_\infty x + \dfrac{Q}{2\pi}\ln r$	$u_\infty y + \dfrac{Q}{2\pi}\varphi$	$u_\infty + \dfrac{Q}{2\pi}\dfrac{x}{x^2+y^2}$	$\dfrac{Q}{2\pi}\dfrac{y}{x^2+y^2}$		

Complex potential	Potential	Stream function	Velocity		Streamlines
$u_\infty\left(z+\dfrac{R^2}{z}\right)$ Parallel flow + dipole = cylinder flow	$u_\infty x\left(1+\dfrac{R^2}{x^2+y^2}\right)$	$u_\infty y\left(1-\dfrac{R^2}{x^2+y^2}\right)$	**On the cylinder:** $2u_\infty\sin^2\varphi$	$-2u_\infty\sin\varphi\cos\varphi$ $2u_\infty\lvert\sin\varphi\rvert$	
$u_\infty\left(z+\dfrac{R^2}{z}\right)+\dfrac{\Gamma}{2\pi}\,\mathrm{i}\ln z$ Cylinder flow + vortex	$u_\infty x\left(1+\dfrac{R^2}{x^2+y^2}\right)-\dfrac{\Gamma}{2\pi}\varphi$	$u_\infty y\left(1-\dfrac{R^2}{x^2+y^2}\right)+\dfrac{\Gamma\ln r}{2\pi}$	**On the cylinder:** $2u_\infty\sin^2\varphi$ $+\dfrac{\Gamma}{2\pi R}\sin\varphi$	$-2u_\infty\sin\varphi\cos\varphi$ $-\dfrac{\Gamma}{2\pi R}\cos\varphi$ $c=\left\lvert 2u_\infty\sin\varphi\right.$ $\left.+\dfrac{\Gamma}{2\pi R}\right\rvert$	
$u_\infty z-\dfrac{\Gamma}{2\pi}\varphi$ Parallel flow + vortex	$u_\infty x-\dfrac{\Gamma}{2\pi}\varphi$	$u_\infty y+\dfrac{\Gamma}{2\pi}\ln r$	$u_\infty+\dfrac{\Gamma}{2\pi}\dfrac{y}{x^2+y^2}$	$\dfrac{\Gamma}{2\pi}\dfrac{x}{x^2+y^2}$	

2. Source-sink flow

$$F(z) = \frac{Q}{2\pi}\ln z = \frac{Q}{2\pi}\left(\ln r + i\varphi\right), \quad z = x + iy = re^{i\varphi},$$

$$\Phi = \frac{Q}{2\pi}\ln r = \frac{Q}{2\pi}\ln\sqrt{x^2 + y^2}, \quad \Psi = \frac{Q}{2\pi}\varphi = \frac{Q}{2\pi}\arctan\frac{y}{x};$$

$$\Phi_x = \frac{Q}{2\pi}\frac{x}{x^2 + y^2}, \quad \Phi_y = \frac{Q}{2\pi}\frac{y}{x^2 + y^2}, \quad c = \sqrt{u^2 + v^2} = \frac{Q}{2\pi r}.$$

Volume flow : $\dot{V} = c \cdot 2\pi r \cdot \ell = Q$ = source or sink strength.

3. Vortex flow

$$F(z) = \frac{\Gamma}{2\pi}i\ln z = \frac{\Gamma}{2\pi}\left(-\varphi + i\ln r\right),$$

$$\Phi = -\frac{\Gamma}{2\pi}\varphi, \quad \Psi = \frac{\Gamma}{2\pi}\ln r; \quad \Phi_x = \frac{\Gamma}{2\pi}\frac{y}{x^2 + y^2}, \quad \Phi_y = -\frac{\Gamma}{2\pi}\frac{x}{x^2 + y^2},$$

$$c = \frac{\Gamma}{2\pi r},$$

completely analogous to the source-sink flow. Γ is called circulation or vorticity and is a measure of the intensity of the rotational motion.

4. Dipole flow

$$F(z) = \frac{m}{z} = \frac{m}{x + iy}\frac{x - iy}{x - iy} = \frac{m(x - iy)}{x^2 + y^2},$$

$$\Phi = \frac{mx}{x^2 + y^2}, \quad \Psi = -\frac{my}{x^2 + y^2}; \quad c = \frac{m}{r^2}.$$

Streamlines : $\Psi = K$ = constant : $x^2 + \left(y + \frac{m}{2K}\right)^2 = \frac{m^2}{4K^2},$

that is, circles come with center on the *y-axis*, all of which pass through the origin. The dipole can be realized by superposition of source and sink, whose distance disappears and intensity simultaneously grows beyond all limits.

5. Superposition of parallel flow with source-sink flow

We treat the case discussed graphically earlier (Fig. 4.50). With horizontal parallel flow is

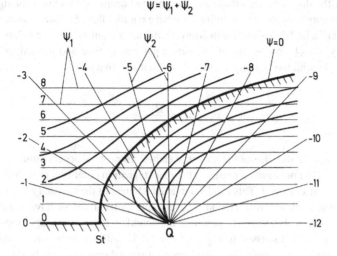

Fig. 4.50 Linear superposition of parallel flow and source

$$F(z) = u_\infty z + \frac{Q}{2\pi}\ln z, \quad \varPhi = u_\infty x + \frac{Q}{2\pi}\ln r, \quad \varPsi = u_\infty y + \frac{Q}{2\pi}\varphi.$$

For the streamlines comes an implicit transcendental equation. Therefore, we have determined their shape graphically above:

$$u = u_\infty + \frac{Q}{2\pi}\frac{x}{x^2 + y^2}, \quad v = \frac{Q}{2\pi}\frac{y}{x^2 + y^2}.$$

For stagnation points thus $c = 0$ and $u = 0$ and $v = 0$. The latter leads to ($y_s = 0$ symmetry to the x-axis), the former then to $x_s = -Q/2\pi u_\infty$. So there is exactly one stagnation point, which lies to the left of the zero point for a source ($Q > 0$) and to the right of it for a sink ($Q < 0$):

$$c^2 = u_\infty^2 + \frac{Qu_\infty}{\pi}\frac{x}{x^2 + y^2} + \frac{Q^2}{4\pi^2}\frac{1}{x^2 + y^2},$$

$$c_p = \frac{\Delta p}{q} = -\frac{Q}{\pi u_\infty}\frac{1}{x^2 + y^2}\left(x + \frac{Q}{4\pi u_\infty}\right).$$

Herewith one can discuss the course of the pressure and velocity distribution on the stagnation streamline. In Fig. 4.51 the case of the source is sketched. In front of the body the pressure increases, the flow is slowed down. At the body, it is accelerated by the displacement effect, and the pressure drops. At this point the incident flow velocity is exceeded. At infinity we have parallel flow. So there comes up the contour of a **half-body obtuse in front**. Its diameter results d_∞ from a balance. The quantity $\dot{V} = Q$ flowing out of the source per unit of time and per unit of depth flows right with the velocity u_∞, thus $\dot{V} = Q = u_\infty d_\infty$, that is,

$$d_\infty = \frac{Q}{u_\infty}. \tag{4.78}$$

In Fig. 4.52 the case of the sink is sketched. Here, flow occurs around a **blunt body tail**. The pressure increases as the rear stagnation point is approached, while the flow is decelerated. This can lead to a detachment of the boundary layer in real flows, that is, flows with friction. The pressure distribution we have determined with the potential theory is imposed on the boundary layer. If we superimpose the individual cases discussed in Figs. 4.51 and 4.52 with the same source and sink strengths, we get a closed body. An important special case will now be discussed.

Fig. 4.51 Source in parallel flow

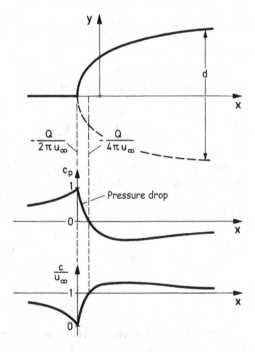

Fig. 4.52 Sink in parallel flow

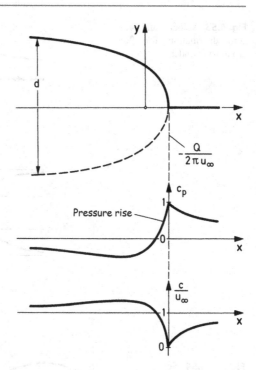

6. Superposition of parallel flow with dipole flow

$$F(z) = u_\infty\left(z + \frac{R^2}{z}\right), \quad \Psi = u_\infty y\left(1 - \frac{R^2}{x^2 + y^2}\right).$$

The stagnation streamline $\Psi = 0$ is given by $y = 0$ and $x^2 + y^2 = R^2$. It is the flow around the cylinder. On the cylinder we obtain: $c = 2u_\infty|\sin \varphi|$, $cp = 1 - 4\sin^2\varphi$. In Fig. 4.53, velocity and pressure on the stagnation streamline are sketched. At the thickness maximum we get the velocity $c_{max}/u_\infty = 2$. A considerable pressure increase occurs on the back side of the body. Due to the symmetry of the pressure distribution in the x and y directions, no resultant force is exerted on the cylinder.

7. Superposition of flow around cylinder and vortex

We go one step further by superimposing a vortex on the example discussed in 6. The scheme in Fig. 4.54 already shows the typical properties of the flow field. There is an **asymmetric** flow with respect to the *x-axis*. The cylinder obviously remains a streamline here as well, since it is a streamline in both subfields. How-

Fig. 4.53 Velocity and pressure distribution for flow around a cylinder

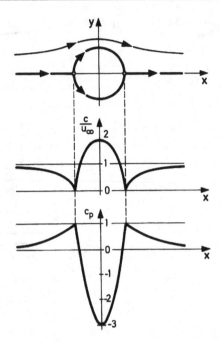

Fig. 4.54 Schematic of the superimposition of cylinder flow and vortex

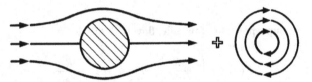

ever, $\Psi = 0$ does not hold on it now, but $\Psi = \Gamma/(2\pi) \ln R$. For velocity and pressure on the cylinder comes

$$c = \left| 2u_\infty \sin\varphi + \frac{\Gamma}{2\pi R} \right|, \; c_p = 1 - \left(\frac{c}{u_\infty} \right)^2 = 1 - \left(2\sin\varphi + \frac{\Gamma}{2\pi u_\infty R} \right)^2. \quad (4.79)$$

The stagnation points are located at

$$\sin\varphi_s = -\frac{\Gamma}{4\pi u_\infty R}.$$

For a clockwise vortex ($\Gamma > 0$), the two stagnation points are located in the third and fourth quadrants. In Fig. 4.55 possible flow fields are sketched. For, $\Gamma = 4\pi u_\infty R$ the two stagnation points coincide ($x = 0, y = -R$) on the contour. If $\Gamma > 4\pi u_\infty R$, then this stagnation point on the *y-axis* moves into the flow field. The flow is in any case symmetrical to the *y-axis*. A force perpendicular to the *x-axis* arises here, a **lift**, the so-called **magnus force**[13]. Figure 4.56 explains the calculation of this force *Fy*:

$$dF_y = -\Delta p \sin\varphi R d\varphi b.$$

b is the width of the cylinder perpendicular to the drawing plane. The sign is chosen so that an overpressure $\Delta p > 0$ causes a downforce d*Fy* < 0.

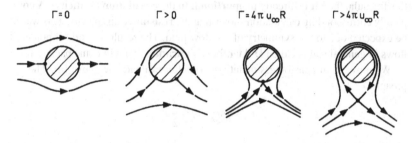

Fig. 4.55 Cylinder flow with circulation

Fig. 4.56 Calculation of the lift force

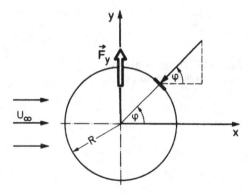

---

[13] H.G. Magnus, 1802–1870.

$$F_y = -bR \int\limits_{\varphi=0}^{2\pi} \Delta p \sin\varphi \, d\varphi$$

$$= -bRq \int\limits_{\varphi=0}^{2\pi} \sin\varphi \left[1 - 4\sin^2\varphi - \frac{2\Gamma}{\pi u_\infty R}\sin\varphi - \frac{\Gamma^2}{4\pi^2 R^2 u_\infty^2} \right] d\varphi$$

$$= \frac{bq2\Gamma}{\pi u_\infty} \int\limits_{\varphi=0}^{2\pi} \sin^2\varphi \, d\varphi = \varrho u_\infty b\Gamma.$$

$$F_y = \varrho u_\infty b\Gamma \tag{4.80}$$

Equation (4.80) is called the **Kutta-Joukowski formula**[14] for lift. According to this formula, the **lift** is directly **proportional to** the **circulation** (vorticity). A completely corresponding calculation shows that the **resistance** disappears. This was to be expected due to the symmetry of the flow field. The result is valid for potential flows in general and is called **D'Alembert's paradox.** We will come back to this.

We move to dimensionless quantities to better evaluate the above lift. The approach

$$F_y = qAc_a, \quad q = \frac{\varrho}{2}u_\infty^2$$

results with the designations of (Fig. 4.57) A = 2bR and the Kutta-Joukowski formula for the lift coefficient

$$c_a = \frac{\Gamma}{u_\infty R}. \tag{4.81}$$

Fig. 4.57 Designations in the case of flow around a cylinder

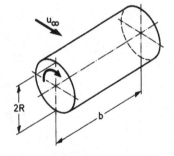

---

[14] W. Kutta, 1867–1944; N.J. Joukowski, 1847–1921.

If we take the limiting case where both stagnation points coincide, $\Gamma = 4\pi u_\infty R$, then (4.81) becomes

$$c_a = 4\pi \approx \pi 12.5. \tag{4.82a}$$

This is an extremely high value compared with results for an airfoil. There they are about one power of ten lower. For example, the following applies to the little inclined (angle α) plane plate

$$c_a = 2\pi \sin \alpha. \tag{4.82b}$$

At $\alpha = 10° \div 0.175$ becomes $c_a \approx 1.1$. Because of the high lift of the rotating cylinder, there has been no lack of attempts to use this effect technically. With the **Flettner rotor**,[15] for example, the transverse force was to be used to propel a ship. Two vertical fast rotating cylinders did drive the ship. Technical difficulties and the occurrence of considerable resistance led to the abandonment of the experiments, although c_a-values of about 9 could be realized.

4.2.5 Potential Flows Around Given Bodies

The examples treated so far serve mainly to gain experience in this field. In the present form they are not yet able to solve the boundary value problem for a **given body**. For this purpose, for example, the **singularity method** is used, which is to be presented in its basic features for slender bodies. Here, **continuous distributions** of singularities (sources/sinks, vortices) are attached to the chord of the profile. The strength of the same is to be dimensioned in such a way that, when superimposed on the parallel flow, the given body contour appears as a streamline. Here it is so that for the symmetrical body in non-adjusted flow (thickness effect) source and sink distributions are used, while for adjustment and curvature vortex assignments are used. In the first case the flow is symmetric about the *x-axis*, while in the second case it is asymmetric. We give the derivation only for the **thickness effect** and refer to the special literature for tilt and kurtosis. The contribution of a differential source-sink element (source point $P_2(\xi,\eta)$) at the receptor point is $P_1(x,y)$ (Fig. 4.58)

$$d\Phi(x,y,\xi,\eta) = \frac{dQ(\xi,\eta)}{2\pi} \ln \sqrt{(x-\xi)^2 + (y-\eta)^2}.$$

---

[15] A. Flettner, 1885–1961.

Fig. 4.58 Source and receptor point

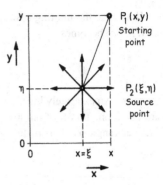

If we only use the chord (= ξ - Axis), the velocities are as follows

$$d\left(u-u_{\infty}\right) = \frac{dQ(\xi)}{2\pi} \frac{x-\xi}{\left(x-\xi\right)^{2}+y^{2}} = \frac{1}{2\pi} \frac{x-\xi}{\left(x-\xi\right)^{2}+y^{2}} \frac{dQ}{d\xi} d\xi.$$

$$dv = \frac{dQ(\xi)}{2\pi} \frac{y}{\left(x-\xi\right)^{2}+y^{2}} = \frac{1}{2\pi} \frac{y}{\left(x-\xi\right)^{2}+y^{2}} \frac{dQ}{d\xi} d\xi.$$

In this $dQ/d\xi$, is the source-sink density. If we occupy the length ℓ, then come to
the integral representations

$$u\left(x,y\right)-u_{\infty} = \frac{1}{2\pi} \int_{0}^{\ell} \frac{x-\xi}{\left(x-\xi\right)^{2}+y^{2}} \frac{dQ}{d\xi} d\xi, \tag{4.83a}$$

$$v\left(x,y\right) = \frac{1}{2\pi} \int_{0}^{\ell} \frac{y}{\left(x-\xi\right)^{2}+y^{2}} \frac{dQ}{d\xi} d\xi. \tag{4.83b}$$

The source-sink density $dQ/d\xi$ is to be determined herein in **such a way** that the
body contour becomes streamline. The condition of the **slender body** leads in the
condition for the streamline (4.7) to the essential simplification

$$\frac{dh}{dx} = \frac{v\left(x,h(x)\right)}{u\left(x,h(x)\right)} \approx \frac{v\left(x,0\right)}{u_{\infty}}. \tag{4.84}$$

The boundary condition is thus satisfied on the profile chord. If we use (4.84) in (4.83b), then with the substitution $\xi - x = ys$, $d\xi = yds$ in the limit $y \to 0$ we get

$$v(x,y) = \frac{1}{2\pi} \int\limits_{\frac{x}{y}}^{\frac{\ell-x}{y}} \frac{dQ(x+ys)}{d\xi} \frac{ds}{1+s^2} \xrightarrow[(y\to 0)]{} \frac{1}{2\pi} \frac{dQ}{dx} \int\limits_{-\infty}^{\infty} \frac{ds}{1+s^2}$$

$$= \frac{1}{2} \frac{dQ}{dx} = u_\infty \frac{dh}{dx},$$

$$\frac{dQ}{dx} = 2u_\infty \frac{dh}{dx}. \tag{4.85}$$

This is a very illustrative result. Sources ($dQ/dx > 0$) have to be placed where the body expands, sinks where it contracts (Fig. 4.59). Equation (4.85), incidentally, also follows immediately from the formula for the thickness of the half-body (4.78), if one sets $d = 2$ h and allows dependence on x *on* the left and right. This then takes into account the effect of the source occupancy instead of the single singularity allowed there. Equation (4.85), together with (4.83a, b), gives the solution to our problem. What remains is a pure integration problem:

$$\frac{u - u_\infty}{u_\infty} = \frac{1}{\pi} \int\limits_0^\ell \frac{(x-\xi)\frac{dh}{d\xi}}{(x-\xi)^2 + y^2} d\xi, \tag{4.86a}$$

$$\frac{v}{u_\infty} = \frac{1}{\pi} \int\limits_0^\ell \frac{y\frac{dh}{d\xi}}{(x-\xi)^2 + y^2} d\xi. \tag{4.86b}$$

Fig. 4.59 Source- sink distribution and body contour

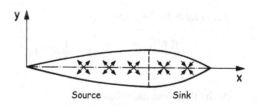

Source   Sink

Fig. 4.60 To calculate Cauchy's main value

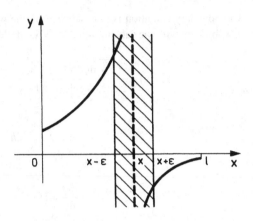

Particularly important is the speed **on** the profile ($y \to 0$). It becomes from (4.86a)

$$\frac{u(x,0)}{u_\infty} = \frac{1}{\pi} \int_0^\ell \frac{\frac{dh}{d\xi}}{x-\xi} d\xi = \lim_{\varepsilon \to 0} \frac{1}{\pi} \left\{ \int_0^{x-\varepsilon} \dots + \int_{x+\varepsilon}^\ell \dots \right\}. \tag{4.87}$$

The singular integral is to be formed herein in terms of Cauchy's principle. Here we exclude the singular point $\xi = x$ symmetrically and proceed to the limit $\varepsilon \to 0$ (Fig. 4.60). We compute the velocity for the parabolic profile:

$$h(x) = 4h_{max} x(1-x) = 2\tau x(1-x), \quad 0 \le x \le 1 \left(\overset{\wedge}{=} \ell \right). \tag{4.88}$$

The thickness parameter of the profile is

$$\tau = \frac{2h_{max}}{\ell}.$$

(4.87) leads to (Fig. 4.61)

$$\frac{u(x,0)-u_\infty}{u_\infty} = \frac{4\tau}{\pi} \left\{ 1 - \left(\frac{1}{2} - x \right) \ln \left| \frac{1-x}{x} \right| \right\}, \quad -\infty < x < +\infty. \tag{4.89}$$

At the thickness maximum is

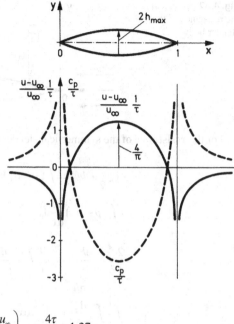

Fig. 4.61 Velocity at parabolic profile

$$\left(\frac{u-u_\infty}{u_\infty}\right)_{max} = \frac{4\tau}{\pi} = 1.27\tau.$$

The velocity distribution shows the characteristics found earlier: in front of the body an accumulation, at the body first acceleration, behind the thickness maximum again deceleration up to the rear stagnation point. In the stagnation points a (weak) logarithmic singularity results as a consequence of our simplified boundary condition. This error influences the flow field only slightly.

The pressure coefficient (4.77) can be simplified for the flow around slender bodies:

$$c^2 = u^2 + v^2 = \left(u_\infty + u - u_\infty\right)^2 + v^2 = u_\infty^2 + 2u_\infty\left(u - u_\infty\right) + \ldots,$$

$$\frac{c^2}{u_\infty^2} = 1 + 2\frac{u - u_\infty}{u_\infty} + \ldots,$$

$$c_p = \frac{\Delta p}{q} = 1 - \frac{c^2}{u_\infty^2} = -2\frac{u - u_\infty}{u_\infty}. \tag{4.90}$$

Fig. 4.62 To calculate
the resistance of the
slender body

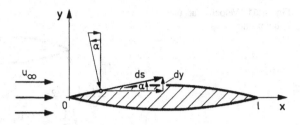

For the **resistance of** the symmetrical slender body we get (Fig. 4.62)

$$W = c_w \frac{\varrho}{2} u_\infty^2 b \ell = 2 \int_0^\ell (p - p_\infty) \sin \alpha b \, ds = 2b$$

$$= \int_0^\ell (p - p_\infty) \frac{dh}{dx} dx, \tag{4.91}$$

$$c_w = \frac{2}{\ell} \int_0^\ell c_p \frac{dh}{dx} dx = -\frac{4}{\ell} \int_0^\ell \frac{u - u_\infty}{u_\infty} \frac{dh}{dx} dx = -\frac{4}{\pi \ell}$$

$$= \int_0^\ell \left(\int_0^\ell \frac{\dfrac{dh}{d\xi}}{x - \xi} d\xi \right) \frac{dh}{dx} dx.$$

The double integral J occurring in this representation is zero. This can be seen immediately by swapping the integration variables and then changing the integration order (Fig. 4.63)

$$J = \int_0^\ell \left(\int_0^\ell \frac{\dfrac{dh}{d\xi}}{x - \xi} d\xi \right) \frac{dh}{dx} dx = (\text{Interchanging the variables}) =$$

$$= \int_0^\ell \left(\int_0^\ell \frac{\dfrac{dh}{dx}}{\xi - x} dx \right) \frac{dh}{d\xi} d\xi = (\text{Change of sequence}) =$$

$$= \int_0^\ell \left(\int_0^\ell \frac{\dfrac{dh}{d\xi}}{\xi - x} d\xi \right) \frac{dh}{dx} dx = -J,$$

$$J = 0.$$

Fig. 4.63 To swap the order of integration

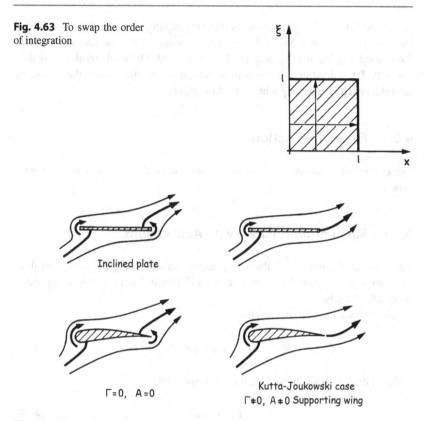

Inclined plate

$\Gamma = 0, \quad A = 0$

Kutta-Joukowski case
$\Gamma \neq 0, \; A \neq 0$ Supporting wing

Fig. 4.64 Flows of the adjusted plate and an airfoil

The resistance is zero. Thus, within the framework of our theory, **D'Alembert's paradox** for potential flows is proven.

If the **flow** is **asymmetrical** to the *x-axis*, which is generated by an **inclination** and/or a **camber**, an additional **vortex occupation of** the chord is necessary. The calculation is similar to the one above, but with the difference that no unique solution results. The total circulation of the profile Γ remains freely selectable here. It is only determined by an additional condition that takes friction into account to some extent. We discuss this qualitatively for the case of the adjusted plate and airfoil (Fig. 4.64). At the beginning of the motion (start-up process), there is an antisymmetric vortex distribution (Fig. 4.64, left). The leading and trailing edges of the plate have flowed around. The total circulation Γ disappears and, according to

(4.80), the lift is zero. Detachment occurs very rapidly at the trailing edge, which leads to the fact that it has no longer flowed around. Then the so-called **Kutta-Joukowski condition** of smooth outflow is fulfilled. The total circulation is thus uniquely determined and different from zero, as is the lift. This is the stationary final state of the supporting wing (Fig. 4.64, right).

4.3   Flow with Friction

The previous considerations serve as preparation for the treatment of flows with losses.

4.3.1   Momentum Theorem with Applications

This general theorem is a balance statement. Numerous experiences from fluid mechanics go into its application. Here we will benefit from the insights gained at special flow fields.

The momentum of a mass element is

$$\mathrm{d}\boldsymbol{J} = \boldsymbol{w}\mathrm{d}m = \varrho\boldsymbol{w}\mathrm{d}V. \qquad (4.92)$$

For a fluid with a volume $V(t)$, the following applies

$$\boldsymbol{J} = \int\limits_{V(t)} \varrho\boldsymbol{w}\mathrm{d}V. \qquad (4.93)$$

The momentum (impulse) theorem reads: **The temporal change of the momentum is equal to the resultant of the external forces.** The external forces $\boldsymbol{F}_{\mathrm{a}}$ are, as usual, the mass and surface forces of the fluid enclosed in the volume V:

$$\frac{\mathrm{d}\boldsymbol{J}}{\mathrm{d}t} = \frac{\mathrm{d}}{\mathrm{d}t}\int\limits_{V(t)} \varrho\boldsymbol{w}\mathrm{d}V = \sum\boldsymbol{F}_{\mathrm{a}}. \qquad (4.94)$$

Now we need the transformation of the time derivative

$$\frac{\mathrm{d}}{\mathrm{d}t}\int\limits_{V(t)} \varrho\boldsymbol{w}\mathrm{d}V.$$

Fig. 4.65 To derive the volume integral

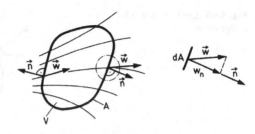

The derivation can be done similar analogous to the continuity (4.64). If we replace the integrand by the scalar function $f(x, y, z, t)$, then (Fig. 4.65) holds

$$\frac{d}{dt}\int_V f dV = \int_V \frac{\partial f}{\partial t} dV + \int_A f(wn) dA. \tag{4.95}$$

The first integral on the right describes the local change of f inside. The second integral gives the resulting flow through the surface. n is the outer normal and is $(wn) dA = d\dot{V}$ the volume flow through the surface element dA. If we replace f by ϱ, then for the **conservation of mass we** get

$$0 = \frac{dM}{dt} = \frac{d}{dt}\int_V \varrho dV = \int_V \frac{\partial \varrho}{\partial t} dV + \int_A \varrho(wn) dA = (\text{Gauss's theorem}) =$$

$$= \int_V \left\{ \frac{\partial \varrho}{\partial t} + div(\varrho w) \right\} dV,$$

that is, (4.65) holds. If we replace f in turn by the components of ϱw and summarize everything, then (4.94) passes into

$$\frac{dJ}{dt} = \frac{d}{dt}\int_V \varrho w dV = \int_V \frac{\partial \varrho w}{\partial t} dV + \int_A \varrho w(wn) dA = \sum F_a. \tag{4.96}$$

The first part describes the local momentum change. This requires knowledge of the flow variables **in the** volume. The second part gives the momentum flow through the surface. Here, the variables occur only **on the surface.**

For **steady** flows, the volume integral is dropped. The flow data are only needed **on** the surface of the **control area.**

Fig. 4.66 Control area with
designations

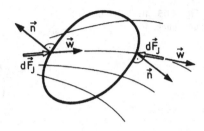

$$\int_A \varrho w (wn) \mathrm{d}A = \sum F_a. \tag{4.97}$$

Let us define as **impulse force** F_J

$$F_J = -\int_A \varrho w (wn) \mathrm{d}A, \tag{4.98}$$

so (4.97) is written in the very simple form

$$F_J + \sum F_a = 0. \tag{4.99}$$

It is true for the impulse (**momentum**) **force** (4.98) that it is locally **parallel** to
w and always directed **into the interior of the control region** (Fig. 4.66), because

$$\mathrm{d}F_J = -\varrho w (wn) \mathrm{d}A. \tag{4.100}$$

If the flow data on **the** boundary are known, it is possible to draw conclusions
about the forces acting on the boundary. When applying this theorem, it is very
important to choose a suitable control area. A lot of experience that we have gained
earlier goes into this. It is important to note that, apart from stationarity, no other
conditions are required. In particular, **lossy flows** are included. The losses enter
here via the boundary conditions. We now treat some **examples**.

4.3.1.1 Flow Through a Manifold

The force exerted by the flowing medium on the inner wall is requested. We assume
that velocity w and pressure p at the inlet (1) and outlet (2) are known (Fig. 4.67).
If we disregard gravity, then (4.99) gives

Fig. 4.67 Curved flow

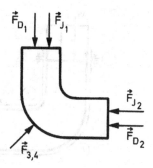

$$0 = \sum F_a + F_J = F_{D_1} + F_{D_2} + F_{3,4} + F_{J_1} + F_{J_2}.$$

With the pressure force

$$F_D = -\int_A p n \, \mathrm{d}A \qquad (4.101)$$

and $F_{3,4}$ as the force which the inner wall of the manifold transmits to the flowing medium. $R = -F_{3,4}$ is the resultant sought, which is exerted by the flow on the inner wall of the manifold. Momentum forces do not occur on the edges (3, 4), since flow does not pass through them. The force map (Fig. 4.68) gives the result by magnitude and direction. In the quantitative treatment, it must be noted that the specifications in the cross-sections (1) and (2) are usually not constant. For momentum and pressure force then the integrals (4.98) as well as (4.101) have to be evaluated. For the velocity at the inlet and outlet Fig. 4.69 shows one possibility. By these specifications the **losses** in the flow field are taken into account.

We deal with the same problem again with a **modified control space.** Let the manifold be free blowing and the control surface be guided along the **outer wall of the manifold**. Thus, this time we get the **total** force transmitted to the manifold. We cut through the bolts holding the flange connection at the inlet. F_B is the holding force we are looking for, assumed arbitrarily at first, which keeps the manifold in equilibrium.

Given: w_1 and p_1 as well as w_2 and $p_2 = p_a$. The control space is shown in Fig. 4.70. If we enter all the forces, we get Fig. 4.71. $F_{D_{3,4}}$ is the pressure force exerted on the manifold shell by the external pressure p_a. The momentum theorem is

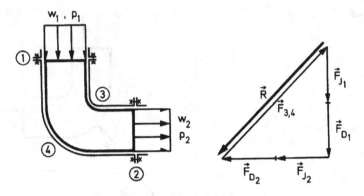

Fig. 4.68 Force map at the manifold

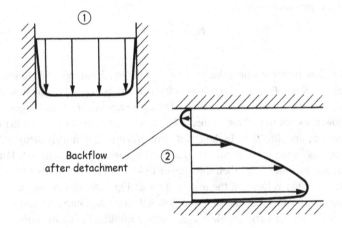

Fig. 4.69 Velocity distribution at inlet and outlet

$$F_{J_1} + F_{J_2} + F_{D_1} + F_{D_2} + F_{D_{3,4}} + F_B = 0.$$

The pressure forces can be simply summarised as follows.

$$F_{D_1} + F_{D_2} + F_{D_{3,4}} = -\left\{ \int_{A_1} (p_1 - p_{-a}) \boldsymbol{n} dA + \int_{A_1} p_{-a} \boldsymbol{n} dA + \int_{A_2} p_a \boldsymbol{n} dA + \int_{A_{3,4}} p_a \boldsymbol{n} dA \right\}$$

$$= -\int_{A_1} (p_1 - p_a) \boldsymbol{n} dA.$$

Fig. 4.70 Control space en-
closes the manifold

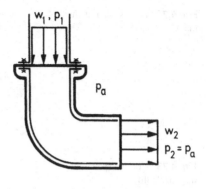

Fig. 4.71 Forces at the control space

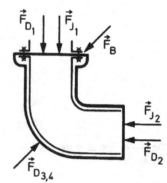

The last three integrals add up to zero, since a constant pressure on a closed surface does not exert any resultant force. If, moreover, velocity and pressure are constant in the respective cross-section, Fig. 4.72 applies.

4.3.1.2 Free-Blowing Nozzle and Diffuser

We are looking for the holding force F_B which acts on the nozzle in the flange connection. Given are velocity w and pressure p each constant over A_1 and A_2, $\varrho = $ constant, moreover $p_2 = p_a$. Figure 4.73 shows an appropriate choice of control space, and Fig. 4.74 shows the applied forces. The direction of F_B is again assumed to be arbitrary; it is determined by the momentum theorem. In the x-direction

$$\varrho w_1^2 A_1 + p_1 A_1 - \varrho w_2^2 A_2 - p_a A_1 + F_B = 0.$$

So with the continuity $w_1 A_1 = w_2 A_2$

Fig. 4.72 Forces map at the manifold

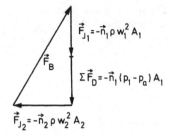

Fig. 4.73 Control space during nozzle flow

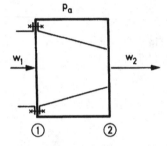

Fig. 4.74 Forces at the control space

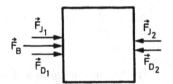

$$F_B = \varrho w_2^2 \left(A_2 - \frac{A_2^2}{A_1} \right) + (p_a - p_1)A_1. \tag{4.102}$$

If we also assume **frictionless** flow here, Bernoulli's equation leads to

$$p_1 - p_a = \frac{\varrho}{2}\left(w_2^2 - w_1^2 \right) = \frac{\varrho}{2} w_2^2 \left(1 - \frac{A_2^2}{A_1^2} \right).$$

Thus (4.102) becomes

Fig. 4.75 Pressure distribution for nozzle and diffuser

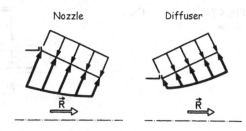

$$F_{\mathrm{B}} = -\frac{\varrho}{2} w_2^2 A_1 \left(1 - \frac{A_2^2}{A_1^2} - 2\frac{A_2}{A_1} + 2\frac{A_2^2}{A_1^2} \right),$$

$$R = -F_{\mathrm{B}} = \frac{\varrho}{2} w_2^2 A_1 \left(1 - \frac{A_2}{A_1} \right)^2 = \frac{\varrho}{2} w_1^2 A_1 \left(\frac{A_1}{A_2} - 1 \right)^2. \tag{4.103}$$

This is the total force transmitted to the nozzle. It acts in the direction of flow, regardless of whether it is ($A_2 < A_1$ nozzle) or ($A_2 > A_1$ diffuser).

The bolts are therefore in any case subjected to tensile stress. This result can be seen immediately on the basis of the pressure distribution (Fig. 4.75). It should be emphasised that this result is only correct under the given conditions. For compressible flows, for example, Laval-nozzle shows quite other results. Here comes up **thrust** which serves for drive. The reader discusses this case.

4.3.1.3 Carnot Shock Diffuser[16]

We consider a diffuser with discontinuous cross-sectional expansion (Fig. 4.76). Space saving issues, among others, lead to such a design. Specifically, there is a complicated flow process. Detachment occurs at the sharp edge followed by mixing. The pressure increase of $1 \to 2$ can be determined without knowing **all the** details of the flow. As preconditions we use:

(a) w_1 and w_2 are constant over the respective cross section.
(b) p_1 is constant over the whole cross section (area A_2) due to detachment.
(c) The wall friction on the inner wall of the diffuser can be neglected.
(d) Stationary flow.

Assumption (a) requires that the diffuser has a length of about 8 diameters. With the results derived later, one can estimate that on such a length the pressure drop due to wall friction is very small.

---

[16] S. Carnot, 1796–1832.

Fig. 4.76 Flow in the Carnot
diffuser

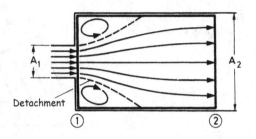

Under these assumptions, the momentum (impulse) theorem for an incompressible flow medium is as follows

$$\varrho w_1^2 A_1 + p_1 A_2 - \varrho w_2^2 A_2 - p_2 A_2 = 0.$$

With continuity

$$\Delta p_{\text{Carnot}} = p_2 - p_1 = \varrho w_1^2 \frac{A_1}{A_2} - \varrho w_2^2 = \varrho w_1 w_2 - \varrho w_2^2,$$

$$\frac{\Delta p_c}{\frac{\varrho}{2} w_1^2} = 2 \frac{w_2}{w_1}\left(1 - \frac{w_2}{w_1}\right) = 2 \frac{A_1}{A_2}\left(1 - \frac{A_1}{A_2}\right). \tag{4.104}$$

With a constant change in cross-section and frictionless flow, the ideal case for the so-called Bernoulli diffuser is

$$\frac{\Delta p_{\text{id}}}{\frac{\varrho}{2} w_1^2} = \frac{p_2 - p_1}{\frac{\varrho}{2} w_1^2} = 1 - \frac{w_2^2}{w_1^2} = 1 - \frac{A_1^2}{A_2^2}. \tag{4.105}$$

The pressure recovery is always greater in the ideal diffuser than in the Carnot diffuser (Fig. 4.77). The difference between the two curves represents a measure of the loss:

$$\frac{\Delta p_{\text{v,c}}}{\frac{\varrho}{2} w_1^2} = \frac{\Delta p_{\text{id}} - \Delta p_c}{\frac{\varrho}{2} w_1^2} = \left(1 - \frac{A_1}{A_2}\right)^2. \tag{4.106}$$

Fig. 4.77 Pressure recovery with the ideal diffuser and with Carnot

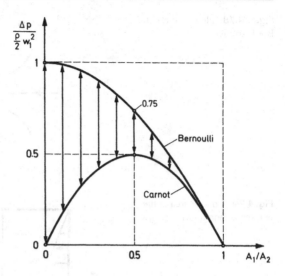

It is particularly significant in the limiting case $A_1/A_2 \rightarrow 0$. Here there is a blow-out into the half-space (A_1 fixed, $A_2 \rightarrow \infty$). With Bernoulli $w_2 = 0$ and $\Delta p \rightarrow (\varrho/2) w_1^2$, that is, the pressure increases by exactly the dynamic pressure. In the Carnot diffuser, on the other hand, there is no pressure increase. The total kinetic energy of the inflow causes a heating of the medium by mixing.

The name **shock diffuser** comes from the analogy to the Carnot shock between two inelastic masses. The loss of kinetic energy there corresponds to the pressure loss here.

4.3.1.4 Borda -orifice[17]

We consider the outflow from a **sharp-edged** orifice (Fig. 4.78). Here there is a jet contraction of $A \rightarrow A_s$, since a jumpy deflection is not possible. The flowing medium creates a rounded outlet for itself, so to speak. The constant pressure p of the environment acts on the free jet boundary. The magnitude of the jet contraction can be determined using the momentum theorem. For this purpose, we consider the outflow from a vessel with a Borda orifice (Fig. 4.79). If we calculate only with the overpressure compared to the atmosphere, an equilibrium occurs in the *x-direction* between the **pressure force** (3.18)

---

[17] J.Ch. de Borda, 1733–1799.

Fig. 4.78 Flow through
Borda -orifice

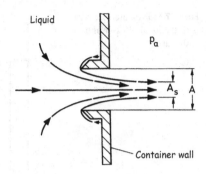

Fig. 4.79 Control area at the
outflow in Borda -orifice

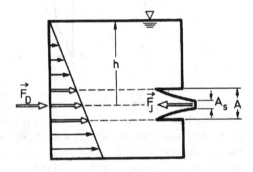

$$F_D = g\varrho h A$$

on the left border and the **impulse (momentum) force** in the beam

$$F_J = \varrho w^2 A_s.$$

If we set both quantities equal, the beam contraction becomes

$$\frac{A_s}{A} = \frac{gh}{w^2}.$$

If we use the Torricellian outflow formula $w = \sqrt{2gh}$ in this case, then

$$\frac{A_s}{A} = \frac{1}{2}. \tag{4.107}$$

The experiments give larger values: 0.5 to 0.6, depending on how far the sharp-edged muzzle projects into the container. The value 0.6 corresponds to the limiting case that the mouth is flush with the container wall.

4.3.1.5 Thrust of an Air-Breathing Engine

We apply the momentum theorem to an aircraft engine. Let the control surfaces be so far away from the engine that the pressure is on them $p = p_\infty$ (Fig. 4.80). The catch cross-section A_∞ is reduced by the propulsion to the jet cross-section A_s with a simultaneous increase in velocity $u_\infty \to u_s$. A **mass flow balance for** the area **outside** the engine provides

$$\varrho_\infty u_\infty \left(A - A_\infty\right) + \dot{m} = \varrho_\infty u_\infty \left(A - A_s\right),$$
$$\dot{m} = \varrho_\infty u_\infty \left(A_\infty - A_s\right).$$

So there is an influx of mass through the control surfaces at the sides. This leads to an impulse force there, whose *x-component* has the value:

$$F_{J,x} = -\int_M \varrho w_x \left(\boldsymbol{wn}\right) \mathrm{d}A = u_\infty \dot{m} = \varrho_\infty u_\infty^2 \left(A_\infty - A_s\right).$$

Thus the momentum theorem is

$$\varrho_\infty u_\infty^2 A + \varrho_\infty u_\infty^2 \left(A_\infty - A_s\right) - \varrho_s u_s^2 A_s - \varrho_\infty u_\infty^2 \left(A - A_s\right) + F_T = 0.$$

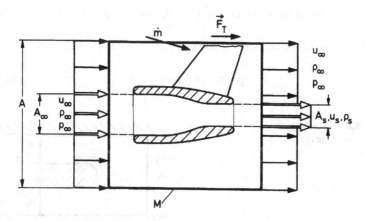

Fig. 4.80 Control surfaces at the power engine

In this F_T is the holding force that occurs at the engine mount so that there is equilibrium. For it comes

$$F_T = \varrho_s u_s^2 A_s - \varrho_\infty u_\infty^2 A_\infty = \dot{m}_T \left(u_s - u_\infty \right) \tag{4.108}$$

with $\dot{m}_T = \varrho_s u_s A_s = \varrho_\infty u_\infty A_\infty$ as mass flow in the engine. For the thrust applies $F_s = -F_T$. It is directly proportional firstly to the mass flow $\dot{m}_T$ and secondly to the increase in velocity $u_s - u_\infty$ in the jet compared to the environment. This makes it clear what possibilities there are for increasing the thrust.

4.3.1.6 Resistance of a Half-Body in the Channel

We investigate the incompressible, frictionless flow around a half-body in a channel (Fig. 4.81). Here there is a **finite** flow cross section, and some differences arise compared to earlier considerations. In particular, the d'Alembert paradox does not apply here. A **resistance** results, which we can determine with the momentum theorem. To determine the effect of force on a body, it is crucial to know what conditions prevail at the **back of** the body. In the case of a half-body, these are not defined a priori. We therefore, cut the cross-section (2), at a sufficient distance from the tip of the body, and assume that in this case is $p = p_2$. Now follow for the basic equations

$$\text{Continuity}: w_1 A_1 = w_2 A_2 \text{ mit } A_2 = A_1 - A. \tag{4.109}$$

$$\text{Momentum theorem}: \varrho w_1^2 A_1 + p_1 A_1 - \varrho w_2^2 A_2 - p_2 A_1 + F_K = 0. \tag{4.110}$$

Here F_K is the holding force of the body and $W = -F_K$ the resistance:

$$W = \varrho w_1^2 A_1 - \varrho w_2^2 A_2 + \left(p_1 - p_2 \right) A_1. \tag{4.111}$$

Fig. 4.81 Flow around a half-body in the channel

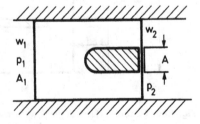

The Bernoulli equation gives for the considered frictionless flow with continuity

$$p_1 - p_2 = \frac{\varrho}{2}\left(w_2^2 - w_1^2\right) = \frac{\varrho}{2}w_1^2\left(\frac{A_1^2}{A_2^2} - 1\right).$$

This leads with (4.111) to

$$W = \frac{\varrho}{2}w_1^2 A_1\left(1 - \frac{A_1}{A_2}\right)^2 = \frac{\varrho}{2}w_1^2 A \frac{\dfrac{A}{A_1}}{\left(1 - \dfrac{A}{A_1}\right)^2}. \tag{4.112}$$

In this we can determine

$$c_w = \frac{\dfrac{A}{A_1}}{\left(1 - \dfrac{A}{A_1}\right)^2} \tag{4.113}$$

as a dimensionless resistance coefficient for a body in the channel. A/A_1 is the characteristic area ratio, which provides a measure of the obstruction. For $A_1/A_2 \rightarrow 1$, that is, in the infinitely extended flow field, goes $W \rightarrow 0$.

4.3.2 Angular Momentum Theorem with Application

Analogous to the impulse theorem, there is a corresponding statement about the moments. This is important for many applications. Because only this results in the points of application of the forces determined above. These considerations are particularly interesting for fluid flow machines. The power absorbed or delivered when flowing through an impeller can be calculated without further ado.

The angular momentum of a mass element is

$$d\boldsymbol{L} = (\boldsymbol{r} \times \boldsymbol{w})dm = \varrho(\boldsymbol{r} \times \boldsymbol{w})dV. \tag{4.114}$$

For a fluid of volume $V(t)$ is therefore

$$L = \int_{V(t)} \varrho(r \times w)\mathrm{d}V. \tag{4.115}$$

The change in time of this angular momentum is equal to the sum of all applied external moments $(= \Sigma\, M_\mathrm{a})$.
The latter result from the mass and surface forces discussed earlier $(= \Sigma\, F_\mathrm{a})$.

$$\frac{\mathrm{d}L}{\mathrm{d}t} = \frac{\mathrm{d}}{\mathrm{d}t}\int_{V(t)} \varrho(r \times w)\mathrm{d}V = \Sigma M_\mathrm{a}. \tag{4.116}$$

Again, we use the relation (4.95) for each component and then summarize everything:

$$\frac{\mathrm{d}L}{\mathrm{d}t} = \frac{\mathrm{d}}{\mathrm{d}t}\int_V \varrho(r \times w)\mathrm{d}V = \int_V \frac{\partial \varrho(r \times w)}{\partial t}\mathrm{d}V + \int_A \varrho(r \times w)(wn)\mathrm{d}A = \Sigma M_\mathrm{a}.$$

The discussion is similar to the case of the momentum theorem. For **steady-state** flows, the volume integral falls away, and we again need the flow data only **on** the surface of the control region:

$$\int_A \varrho(r \times w)(wn)\mathrm{d}A = \Sigma M_\mathrm{a}. \tag{4.117}$$

This prerequisite is somewhat problematic this time. A fluid flow engine is predominantly connected with an unsteady flow. Only in a system rotating with the impeller one can speak of a steady flow. Let us define as angular **momentum M_J** analogous to (4.98):

$$M_J = -\int_A \varrho(r \times w)(wn)\mathrm{d}A, \tag{4.118}$$

then (4.117) is written similarly to the impulse theorem in the very simple form

$$M_J + \Sigma M_\mathrm{a} = 0. \tag{4.119}$$

It holds for the angular **momentum** (4.118) that it is locally **parallel to $r \times w$**, because

$$dM_J = -\varrho(r \times w)(wn)dA. \tag{4.120}$$

The meaning is explained with an example.

4.3.2.1 Flow Through a Radial Impeller

A turbine runner (angular velocity ω) is flown radially from the outside to the inside (Fig. 4.82). The fluid enters at r_1 with absolute velocity $c_1 = (c_{1r}, c_{1u})$ into the impeller channel and leaves it at $r_2 < r_1$ with the absolute velocity $c_2 = (c_{2r}, c_{2u})$. If we place the control region so that it coincides with a blade channel (Fig. 4.83), then follows

Fig. 4.82 Flow through a turbine runner

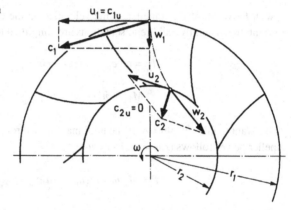

Fig. 4.83 Control area in the impeller channel

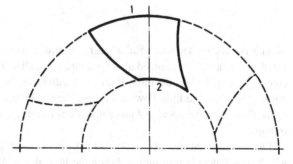

$$0 = \sum M_a + M_J = M_s + M_{J_1} + M_{J_2}.$$

The pressure forces at the inlet and outlet give no momentum, as they are directed radially. M_s is the angular momentum transferred from the blades (that is, from outside!) to the fluid. $-M_s = M_{Tu}$ is then the usable turbine momentum to be removed at the shaft. With $w = c_1$ resp. $c2$ becomes

$$M_{J_1} = \int_\varphi \varrho r_1^2 c_{1u} c_{1r} \mathrm{d}\varphi b,$$

$$M_{J_2} = -\int_\varphi \varrho r_2^2 c_{2u} c_{2r} \mathrm{d}\varphi b,$$

with $b=$ width of the impeller. If the velocities on the corresponding radii are constant, the generally valid representations are simplified to.

$$M_{J_1} = \dot{m} r_1 c_{1u}, \quad M_{J_2} = -\dot{m} r_2 c_{2u},$$

$$M_{Tu} = \dot{m} \left(r_1 c_{1u} - r_2 c_{2u} \right). \tag{4.121}$$

We want to take $\dot{m}$ already as the total mass flow through the impeller. For the impeller power follows ($r_1 \omega = u_1$, $r_2 \omega = u_2$)

$$P = M_{Tu}\omega = \dot{m} \left(u_1 c_{1u} - u_2 c_{2u} \right). \tag{4.122}$$

The (specific) work delivered per unit mass is

$$\frac{P}{\dot{m}} = u_1 c_{1u} - u_2 c_{2u}. \tag{4.123}$$

This equation is known as **Euler's turbine equation.** It is valid with sign reversal of the right side unchanged also for a pump impeller. Then (4.123) is the specific work absorbed by the fluid. Both cases differ only in the direction of energy transfer between the fluid flowing through the machine and the rotating components. One often also speaks of **force machines** (turbines) and **working machines** (pumps).

Finally, a special case of (4.121) should be pointed out. If there is an impeller to which no moment is transmitted during the flow, that is $M_{Tu} = 0$, then obviously either $\dot{m} = 0$ or $r c_u = $ constant holds. The first case is trivial, in the second case the circumferential component is that of a potential vortex. The flow is on spiral paths (Sect. 4.1.3).

4.3.3   Basics of the Influence of Friction: Similarity Parameters

With the momentum theorem, as we have seen, frictional losses can be taken into account. They enter in a roundabout way via the specifications on the boundary of the control space. These can be taken from measurements or calculations. We want to determine the friction influence quantitatively in a particularly simple case. Essentially, this is an extension of the flow filament theory discussed earlier; s and n denote the coordinates in the direction of flow and perpendicular to it (Fig. 4.84). Only the influence of the shear stress τ acting tangentially to the flow direction will be investigated here. A friction tensor occurs in the complete consideration. We will come back to this in Sect. 4.3.10. The individual layers are interconnected by internal friction. This is the cause of the shear stress τ. We cut out a mass element and choose an **arbitrary** velocity profile in the *n-direction* (Fig. 4.84, right). The arrows indicate the directions of the frictional forces transmitted **externally** to the element under consideration (Fig. 4.84, left). They depend on the chosen profile. A small difficulty arises here in choosing the correct sign, which is why this example is discussed in detail. For the case outlined in Fig. 4.84, the following occurs

$$\frac{\text{Frictional force}}{\text{Mass}} = \frac{dR}{dm} = \frac{\left\{-\left(|\tau| + d|\tau|\right) + |\tau|\right\} ds\,db}{\varrho\,ds\,dn\,db} = -\frac{1}{\varrho}\frac{d|\tau|}{dn}, \qquad (4.124)$$

Fig. 4.84 On the friction influence in the flow filament

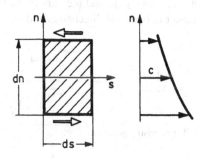

where the magnitude of the shear stress $|\tau|$ using the Newtonian approach is given by

$$|\tau| = \begin{cases} \eta \dfrac{\partial c}{\partial n}, & \dfrac{\partial c}{\partial n} > 0, \\[2mm] -\eta \dfrac{\partial c}{\partial n}, & \dfrac{\partial c}{\partial n} < 0. \end{cases} \tag{4.125}$$

If we enter (4.125) into (4.124) for the case shown in Fig. 4.84, then

$$\frac{dR}{dm} = \frac{1}{\varrho} \frac{\partial}{\partial n}\left(\eta \frac{\partial c}{\partial n}\right). \tag{4.126}$$

If one chooses a different velocity profile in Fig. 4.84, for example, with $\partial c/\partial n > 0$, then again (4.126) is obtained. In the special case η = constant (4.126) simplifies to

$$\frac{dR}{dm} = \frac{\eta}{\varrho} \frac{\partial^2 c}{\partial n^2} = v \frac{\partial^2 c}{\partial n^2} \cdot s \tag{4.127}$$

The friction force therefore, depends on the second derivative of the velocity. The reason for this is obviously that it is the **change in** shear stress perpendicular to the flow filament is important. In Couette flow, a particle therefore, experiences no resulting frictional force.

We now compile some forces acting on a mass element. These are typical representatives of the corresponding influences. In the bottom line the individual terms are represented by characteristic reference quantities for time (t), length (ℓ), velocity (c), density (ϱ) and pressure (p) of the flow filament. We use the same length scale ℓ in both axis directions s and n .

Physical effect	Inertia		Pressure	Gravity	Friction
	a	b			
Force/Mass	$\dfrac{\partial c}{\partial t}$	$c\dfrac{\partial c}{\partial s}$	$\dfrac{1}{\varrho}\dfrac{\partial p}{\partial s}$	$g\dfrac{\partial z}{\partial s}$	$v\dfrac{\partial^2 c}{\partial n^2}$
Characteristic quantities	$\dfrac{c}{t}$	$\dfrac{c^2}{\ell}$	$\dfrac{p}{\varrho\ell}$	g	$\dfrac{vc}{\ell^2}$

From these five typical forces, four independent dimensionless force (= ratios) can be formed. These **ratios** characterize a flow field and describe the incoming physical effects. We obtain in turn:

1. $\dfrac{\text{Pressure force}}{\text{Inertial force }(b)} \sim \dfrac{\dfrac{p}{\varrho\ell}}{\dfrac{c^2}{\ell}} = \dfrac{p}{\varrho c^2} = \text{Euler-or Newton-number} = \text{Eu} = \text{Ne}.$ (4.128)

For a compressible medium

$$\text{Eu} = \frac{p}{\varrho c^2} = \frac{\kappa p}{\varrho}\frac{1}{c^2}\frac{1}{\kappa} = \frac{1}{\kappa M^2},$$

which results in a connection with the Mach number. We have already repeatedly encountered Euler numbers. We recall, for example, the pressure coefficient (4.77).

2. $\dfrac{\text{Inertial force}(b)}{\text{Gravity}} \sim \dfrac{c^2}{\ell g} = \text{Froude-Number} = \text{Fr.}$ (4.129)

The Froude number[18] is important wherever gravity has a significant influence on the flow, for example, in waters with a free surface.

3. $\dfrac{\text{Inertial force}(a)}{\text{Inertial force}(b)} \sim \dfrac{\ell}{tc} = \text{Strouhal-number} = \text{Str.}$ (4.130)

---

[18] W. Froude, 1810–1879.

This ratio characterizes **unsteady** flow processes, as they occur, for example, in all periodically operating force machines and working machines. The Strouhal number[19] is obtained by relating the transient term of the Bernoulli eq. (4.13) to the steady-state terms. This is often necessary to determine whether a flow can be considered steady-state. For this purpose, it must be Str $\ll 1$.

$$\textbf{4.} \quad \frac{\text{Inertial force}(b)}{\text{Frictional force}} \sim \frac{\dfrac{c^2}{\ell}}{\dfrac{vc}{\ell^2}} = \frac{c\ell}{v} = \text{Reynolds-Number} = \text{Re.} \qquad (4.131)$$

This important ratio (similarity parameter) measures the influence of friction.

$$\text{Re} = \frac{c\ell}{v} \gg 1, \qquad (4.131a)$$

that is, if the inertial force (b) is much greater than the friction force, the friction **within the** flow field has little influence. Viscosity only plays a role near the wall due to the adhesion condition in the boundary layer (Fig. 4.85). This is the starting point of Prandtl's boundary layer theory. If

$$\text{Re} = \frac{c\ell}{v} < 1, \qquad (4.131b)$$

the friction in the whole flow field is of importance. By Re < 1 so called creeping flows (= Stokes flows[20]) are captured. Here the non-linear inertia terms in the equations of motion can often be completely neglected. The pressure forces are in equilibrium with the friction forces. Examples are the motions in very viscous oils. However, (4.131b) is also important for flows at extremely low density, because the Reynolds number

$$\text{Re} = \frac{\varrho c\ell}{\eta}$$

---

[19] V. Strouhal, 1850–1922.
[20] G.G. Stokes, 1819–1903.

Fig. 4.85 Velocity profile in
the boundary layer

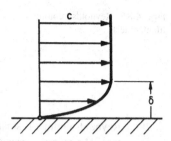

becomes small with ϱ. These considerations play a role for example, for satellite movements at the edge of the atmosphere, and in the laboratory for example, for vacuum pumps.

Difficulties can sometimes arise in choosing the appropriate reference quantities for the similarity parameters. Here, a certain experience is needed to find out those determinants which are crucial for the physics of the respective flow. Sometimes there are several possibilities. For example, in the Re number of boundary layer flow (Fig. 4.85), the boundary layer thickness δ can be used as the length measure. Similarly, one can also use the run length ℓ, from the tip of the body to the point under consideration. Both formations are useful, although they lead to different orders of magnitude of the Re number.

The most important application of the ratios (similarity parameters) is that with their help it is possible to convert **geometrically similar flow fields** into each other. This is the basis of all **model tests** were, on the basis of measurements on a geometrically similarly reduced model, for example, in a wind tunnel or water channel, statements are to be made about the large-scale design. We will come back to this in Sect. 4.3.16.

4.3.4 Laminar and Turbulent Flow

We observe two fundamentally different flow states in the experiment. They were described qualitatively by **Hagen**, and quantitatively recorded for the first time by **Reynolds**. The experimental facts are explained by means of Reynolds' coloured filament experiment (Fig. 4.86). A viscous medium (kinematic viscosity ν) flows through a tube of circular cross-section (diameter d) with velocity c. The flow rate can be changed using a throttle. A coloured filament is used as a flow indicator.

Fig. 4.86 Reynolds colour
filament test

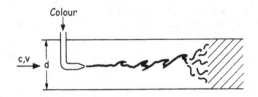

1. If the Reynolds number is small, that is, Re = cd/v < 2300, then a **laminar** flow
   is present. The macroscopically observable flow takes place in parallel layers
   (lamina = layer, disc). Microscopically, that is, molecularly, a random exchange
   of momentum takes place between the individual layers, which, as we stated
   earlier, is the cause of the internal friction.
2. If the Reynolds number is large, i.e., Re = cd/v > 2300, then one speaks of **turbulent** flow. Here, in contrast to the above, a macroscopic, visible exchange
   occurs. It is an unsteady, vortex-like random motion.

Reynolds studied the transition from laminar to turbulent flow (so-called laminar-turbulent transition) and found that it depends solely on the cd/v ratio. Based on
observations, he suspected a **stability problem**. The laminar flow becomes unstable to perturbations at higher Reynolds numbers, that is, small perturbations, which
are always present in nature and technology, cause large effects in such a case,
which finally transform the laminar flow into the turbulent flow.

This concrete recording of the turbulent state leads to the **Reynolds description
of turbulent flows**. Here, the unsteady field variable (for example, the velocity

$u(x, y, z, t)$) is additively decomposed into a temporal mean value $\bar{u}(x,y,z)$ and a
fluctuation variable $u'(x, y, z, t)$.

$$u(x,y,z,t) = \bar{u}(x,y,z) + u'(x,y,z,t), \qquad (4.132)$$

$$v = \bar{v} + v', \quad w = \bar{w} + w'.$$

The time average at the fixed location is defined by

$$\bar{u}(x,y,z) = \frac{1}{T}\int_0^T u(x,y,z,t)\,\mathrm{d}t. \qquad (4.133)$$

Fig. 4.87 Hot-wire signal as a function of time

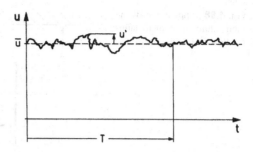

T is chosen in this to be so large that a further increase does not result in an appreciable change in $\bar{u}$. (4.133) has the consequence that the time averages of the fluctuation variables disappear:

$$\bar{u}' = \bar{v}' = \bar{w}' = 0. \tag{4.134}$$

These fluctuation velocities $\bar{u}' = \bar{v}' = \bar{w}' = 0$, which contain the characteristic properties of turbulent flows, can be determined with a **hot-wire probe**. Here the cooling of a heated platinum wire is used as a measure of the fluctuations. In a fixed point in space (x, y, z), a representation as in Fig. 4.87 is obtained.

To characterize the **degree of turbulence** (=Tu) in a flow field, the following dimensionless formation is used

$$\mathrm{Tu} = \frac{\sqrt{\overline{(u')^2}}}{\bar{u}}. \tag{4.135}$$

The numerator contains the root of the mean square error as a characteristic measure of the fluctuation variable. It is set in relation to the mean flow velocity at the point under consideration.

4.3.5 Velocity Distribution and Pressure Drop in Circular Pipes with Laminar and Turbulent Flow

1. Laminar pipe flow (Hagen-Poiseuille flow)

We consider a horizontal pipe section. Let the flow be **developed**, that is, the velocity profile does not change in the *x-direction*. This assumes that we are at a sufficiently large distance from the inlet. We will return to an estimate of this dis-

Fig. 4.88 Application of the momentum theorem to laminar flow in a circular pipe

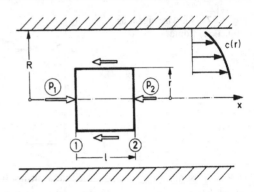

tance later. For a layered flow in the pipe, the pressure is constant across the cross section, as in the boundary layer. Consider (4.16) in the case $r \to \infty$. A pressure difference in **the** direction of flow maintains the motion. To determine the velocity profile, we apply the momentum theorem to a coaxial cylinder (Fig. 4.88). There is an equilibrium between the pressure forces and the friction force. A resultant momentum force does not enter here because the flow is developed. We obtain for the case sketched in Fig. 4.88

$$\pi r^2 p_1 - \pi r^2 p_2 - |\tau| 2\pi r \ell = 0.$$

The following applies analogously to (4.125)

$$|\tau| = \begin{cases} -\eta \dfrac{dc}{dr}, & \dfrac{dc}{dr} < 0 \\[2ex] \eta \dfrac{dc}{dr}, & \dfrac{dc}{dr} > 0 \end{cases}$$

$$|\tau| = (p_1 - p_2)\frac{r}{2\ell} = \frac{\Delta p}{2\ell} r = -\eta \frac{dc}{dr}. \tag{4.136}$$

The shear stress is thus a **linear** function of r. If we assume a different velocity distribution in Fig. 4.88, the same relationship as above is obtained, namely

$$\frac{dc}{dr} = -\frac{\Delta p}{\ell} \frac{1}{2\eta} r.$$

Fig. 4.89 Velocity and shear stress for laminar flow in a circular pipe

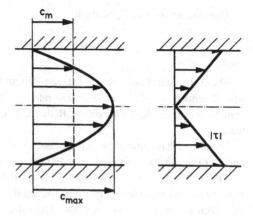

Integration provides with the adhesion condition ($r = R$, $c = 0$)

$$c(r) = \frac{\Delta p}{\ell}\frac{R^2}{4\eta}\left(1 - \frac{r^2}{R^2}\right) = c_{max}\left(1 - \frac{r^2}{R^2}\right).$$ (4.137)

A parabolic velocity distribution $c(r)$ results with the maximum velocity

$$c_{max} = \frac{\Delta p}{\ell}\frac{R^2}{4\eta}$$ (4.138)

on the axis of rotation (Fig. 4.89). By integration, the volume flow

$$\dot{V} = c_m A = \int_A c\,dA = \int_{r=0}^{R} c_{max}\left(1 - \frac{r^2}{R^2}\right)2\pi r\,dr = \pi R^2 \frac{c_{max}}{2} = A\frac{c_{max}}{2}.$$ (4.139)

Thus, the volumetric mean of the velocity c_m follows

$$c_m = \frac{1}{2}c_{max}.$$ (4.140)

This results in the representation for the volume flow

$$\dot{V} = c_m A = \frac{1}{2}c_{max}A = \frac{\pi}{8}\frac{\Delta p R^4}{\ell \eta}.$$

Thus the proportionalities result

$$\dot{V} \sim \Delta p, \quad \dot{V} \sim R^4. \tag{4.141}$$

These statements are called the Hagen-Poiseuille law. They show the characteristic dependencies of the volume flow rate. $\dot{V} \sim R^4$ is particularly important for applications in the field of medicine. Reduction of R can lead $\dot{V}$ to a drastic reduction.

So far, we have investigated the question of what velocity arises as a consequence of the pressure difference Δp. For the applications, the reversal of the question is of interest: How large is the **pressure decrease** (= pressure loss) Δp in a pipeline for a given volume flow? The frictional influence is expressed in this pressure decrease in the direction of flow. The velocity profile remains unchanged. If we combine (4.138) and (4.140), we get

$$\Delta p = \frac{4\eta \ell c_{\max}}{R^2} = \frac{8 \varrho v \ell c_{\mathrm{m}}}{R^2}.$$

We split this expression by combining characteristic quantities:

$$\Delta p = \frac{\varrho}{2} c_{\mathrm{m}}^2 \frac{\ell}{D} \lambda_{\mathrm{lam}}, \quad \lambda_{\mathrm{lam}} = \frac{64}{\mathrm{Re}_D}, \quad \mathrm{Re}_D = \frac{c_{\mathrm{m}} D}{v}. \tag{4.142}$$

The first part on the right side provides the dimension of the pressure, the second part characterizes the geometry, the third contains the physics of the pipe friction process. λ is called the loss coefficient. This structure of the pressure loss formula is typical. We will encounter it repeatedly. (4.142) contains the other interesting statements:

$$\Delta p \sim \ell, \quad \Delta p \sim c_{\mathrm{m}}. \tag{4.143}$$

The pressure drop is a linear function of the pipe length. This is plausible, since no pipe section is distinguished when the flow is **developed.** Therefore, only a **linear** function is possible. $\Delta p \sim c_{\mathrm{m}}$ is typical for **laminar** flows.

In the applications are given, for example: $\dot{V}$, A, ℓ ϱ, v. From $\dot{V} = c_{\mathrm{m}} A$ follows $c_{\mathrm{m}} = \dot{V} / A$ and hence $\mathrm{Re}_D = (c_{\mathrm{m}} / v)\sqrt{4A / \pi}$. Hereby we check whether is $\mathrm{Re}D \gtrless 2300$. If the laminar case is present, the pressure drop can be calculated using (4.142). In practice, however, we are largely dealing with turbulent flows, which is why we now deal with this case in detail.

2. Turbulent pipe flow

Turbulent flow is much more difficult to treat than laminar flow. In technical applications, often not all details of the flow field are of interest. It is often sufficient to know the time-averaged **mean values.** We assume that there is a developed flow in the pipe. We extend the control space over the entire pipe cross-section (Fig. 4.90). There is an equilibrium between pressure and wall shear stress forces.

$$\pi R^2 \overline{p}_1 - \pi R^2 \overline{p}_2 - \mid \overline{\tau}_w \mid 2\pi R \ell = 0,$$

$$\Delta \overline{p} = \overline{p}_1 - \overline{p}_2 = \mid \overline{\tau}_w \mid \frac{2\ell}{R}. \tag{4.144}$$

For the wall shear stress, we make the dimensional approach

$$\mid \overline{\tau}_w \mid = \frac{\varrho}{2} \overline{c}_m^2 \sigma. \tag{4.145}$$

Herein means the $\overline{c}_m$ temporal and spatial mean value of the velocity. It is obtained from the function $c(r,t)$, by successively carrying out the meanings (4.133) and (4.139). If one enters (4.145) into (4.144), then with the abbreviation $4\sigma = \lambda_{turb}$

$$\Delta \overline{p} = \frac{\varrho}{2} \overline{c}_m^2 \frac{\ell}{D} \lambda_{turb}. \tag{4.146}$$

Fig. 4.90 Control area for turbulent flow in the circular pipe

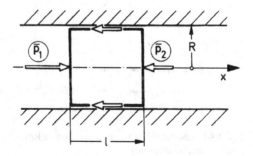

The structure of the pressure loss formula is the same as in the laminar case (4.142). However, this time λ_{turb} must be determined from experiments. A theoretical (analytical) calculation is not possible so far. One obtains.

1. by interpolation of measurement results (**Blasius formula**[21])

$$\lambda_{turb} = \frac{0.3164}{Re_D^{1/4}}, \quad \text{valid until } Re_D \approx 10^5, \tag{4.147a}$$

2. implicit representation of **Prandtl**

$$\frac{1}{\sqrt{\lambda_{turb}}} = 2\log^{10}\left(Re_D \sqrt{\lambda_{turb}}\right) - 0.8, \quad \text{valid until} Re_D \approx 3 \cdot 10^6. \tag{4.147b}$$

In Fig. 4.91 the laminar law and the above relations are plotted. In addition, one can also find the results for **rough pipes**, which will be discussed in detail in the next section. In this context, one speaks of the **Nikuradse diagram**[22]. R/k_s is called

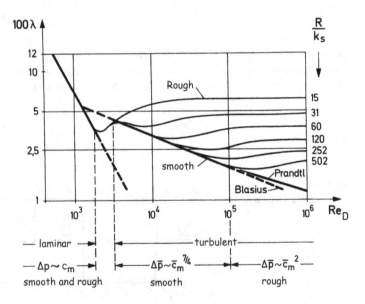

Fig. 4.91 Loss coefficient λ as a function of Reynolds number and roughness for circular pipe (Nikuradse diagram)

[21] H. Blasius, 1883–1970.

[22] J. Nikuradse, 1894–1979.

Fig. 4.92 Definition of sand grain roughness k_s

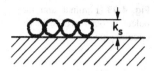

sand grain roughness parameter, k_s is a typical measure of sand grain roughness (Fig. 4.92). Increasing roughness, that is, decreasing R/k_s, means increasing pressure loss. We will return to this in detail.

From **Blasius' law** follows an interesting consequence for the (time-averaged) **velocity**. We enter (4.147a) into (4.145) and refrain from numerical factors:

$$|\bar{\tau}_w| = \frac{\varrho}{2}\bar{c}_m^2 \frac{\lambda_{turb}}{4} \sim \varrho \bar{c}_m^{7/4} v^{1/4} R^{-1/4} \sim \varrho \bar{c}_{max}^{7/4} v^{1/4} R^{-1/4}. \tag{4.148}$$

We use a power approach with free exponent m for the velocity profile:

$$\bar{c}(y) = \bar{c}_{max}\left(\frac{y}{R}\right)^m. \tag{4.149}$$

y is the wall-normal distance $y = R - r$. We solve (4.149) for $\bar{c}_{max}$ and enter this into (4.148):

$$|\bar{\tau}_w| \sim \varrho \bar{c}^{7/4} y^{-7m/4} v^{1/4} R^{7m/4-1/4}. \tag{4.150}$$

Prandtl and v. Kármán[23] have expressed the hypothesis that this turbulent wall shear stress should be independent of the pipe radius. That is, the turbulent flow is more or less determined by the local data of the flow field. In the above case this means

$$m = \frac{1}{7}, \quad \bar{c} = \bar{c}_{max}\left(\frac{y}{R}\right)^{1/7} = \bar{c}_{max}\left(1 - \frac{r}{R}\right)^{1/7}. \tag{4.151}$$

This is the important $1/7$-power law whose validity coincides with that of Blasius' law.

In Fig. 4.93, the laminar (4.137) and turbulent velocity profiles (4.151) are drawn for the same volume flow. The turbulent profile is more rectangular than the laminar one. We discuss some statements in detail.

---

[23] Th. v. Kármán, 1881–1963.

Fig. 4.93 Laminar and turbulent velocity profile in the
circular pipe

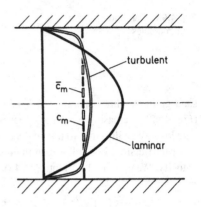

1. For $m = 1/7$ is $\overline{c}_m = 0.816\overline{c}_{max}$. The velocity profile (4.151) has two small flaws. An infinite slope results at the pipe wall. This is not a concern because in the immediate vicinity of the wall the flow is laminar (frictional sublayer) and therefore the above law is not needed there. A bend in the velocity profile occurs at the pipe axis.
2. As the Reynolds number increases, the exponent in (4.151) becomes smaller, that is, the profile becomes more and more rectangular. The reason for this is that the macroscopic transverse exchange strives to balance the velocity profile and make it as uniform as possible across the cross-section.

4.3.6   Laminar and Turbulent Flow Through Rough Pipes (Nikuradse Diagram)

We discuss Fig. 4.91 in detail. The essence can be summarized in two points.

1. For **laminar** flow is $\lambda = f(\text{Re})$, that is, the pressure loss coefficient does not depend on the roughness.
2. For **turbulent** flows the alternative applies

(a) $\lambda = g\left(\text{Re}, \dfrac{R}{k_s}\right)$ for $2 \cdot 10^3 < \text{Re} < 3 \cdot 10^5$,

(b) $\lambda = h\left(\dfrac{R}{k_s}\right)$ for $3 \cdot 10^5 < \text{Re}$.

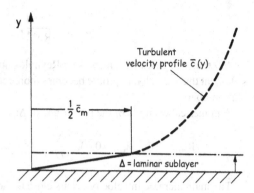

Fig. 4.94 Definition of the thickness of the laminar sublayer Δ

In the range of medium Reynolds numbers ($2 \cdot 10^3 < $ Re $ < 3 \cdot 10^5$), both arguments occur, while for higher Reynolds numbers only the roughness enters. These characteristic dependencies can be explained as follows.

1. In **laminar** flow, there is no significant influence of the roughness, since the macroscopic cross exchange is missing. The laminar flow creates a smooth wall for itself, so to speak, and covers the roughness.
2. **2a, b.** In **turbulent** flow, the decisive factor is whether the roughnesses are still covered by the laminar lower layer near the wall—in which case the pipe is called **hydraulically smooth**—or whether they protrude from this layer and thus significantly influence the fully turbulent flow.

We want to confirm this important idea quantitatively. To do this, we estimate the thickness of the laminar sublayer Δ. In it, the velocity should increase linearly from zero to $1/2 \, \bar{c}_m$ approximately (Fig. 4.94). The wall shear stress can be represented in two ways:

$$|\bar{\tau}_w| = \frac{\varrho}{2} \bar{c}_m^2 \frac{\lambda_{\text{turb}}}{4} = \eta \left(\frac{d\bar{c}}{dy} \right)_w = \varrho \nu \frac{\frac{1}{2} \bar{c}_m}{\Delta}.$$

$$\frac{\Delta}{D} = \frac{4}{\lambda_{\text{turb}}} \frac{1}{\text{Re}_D}. \tag{4.152}$$

If we use the Blasius formula (4.147a) in this, then

$$\frac{\Delta}{D} = \frac{12.64}{\text{Re}_D^{3/4}}, \tag{4.153}$$

Δ/D thus decreases with increasing Reynolds number. This confirms the earlier statement that the velocity profile becomes more rectangular with increasing Reynolds number.

A numerical example of the magnitude of Δ:

$$\text{Re}_D = 10^4, \quad \lambda_{\text{turb}} = 0.03, \quad \frac{\Delta}{D} \approx 10^{-2}, \quad D = 10\text{cm}, \quad \Delta \approx 1\text{mm}.$$

The main increase in velocity occurs close to the wall, that is, over a distance of about 1% of the diameter.

Using the estimate (4.152), we can easily justify the above findings 2a, b. We consider two typical numerical examples:

1. $\text{Re}_D = 10^5$, $\lambda_{\text{turb}} = 0.04$, $\frac{R}{k_s} = 30$

Here we show that the roughnesses protrude from the sublayer. With (4.152) namely

$$\frac{R}{\Delta} = \frac{\lambda_{\text{turb}} \text{Re}_D}{8} = \frac{0.04 \cdot 10^5}{8} = 500 \gg \frac{R}{k_s} = 30, \quad \text{that is } k_s \gg \Delta.$$

2. $\text{Re}_D = 10^4$, $\lambda_{\text{turb}} = 0.03$, $\frac{R}{k_s} = 60$

Now comes

$$\frac{R}{\Delta} = \frac{0.03 \cdot 10^4}{8} = 37.5 < \frac{R}{k_s}, \quad \text{that is } k_s < \Delta.$$

This means that the roughnesses are covered by the underlayer. The pipe is hydraulically smooth.

The transition to **technically rough pipes** is made possible by the concept of **equivalent sand grain roughness**. This is understood to be the sand grain roughness ks which provides the same loss coefficient λ for the same Reynolds number. Figure 4.95 shows some roughness values for technically important cases.

If a pipe of **non-circular cross-section is** present, the **hydraulic diameter** D_h is used instead of D as the characteristic length dimension in the **turbulent** case:

$$D_h = \frac{4A}{U} \tag{4.154}$$

Fig. 4.95 Equivalent sand grain roughnesses

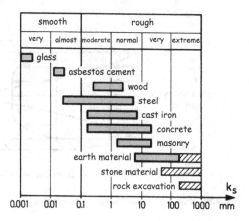

with A= cross-sectional area and U= wetted perimeter. For the pressure loss applies

$$\Delta\bar{p} = \frac{\varrho}{2}\bar{c}_\mathrm{m}^2\frac{\ell}{D_\mathrm{h}}\lambda_\mathrm{turb}, \quad \mathrm{Re}_{D_\mathrm{h}} = \frac{\bar{c}_\mathrm{m}D_\mathrm{h}}{\nu}. \quad (4.155)$$

Note that this statement is limited to turbulent flows. There, the velocity is almost constant over the cross-section. This is probably the reason that in these cases a conversion to the hydraulic diameter is possible. This does not apply to laminar flows, and the pressure loss must then be calculated or measured.

4.3.7 Flow in the Inlet Section

So far we have dealt with the developed flow in the pipe. We now discuss the conditions at the pipe inlet.

1. **Laminar flow**

The fluid is sucked in from the resting state (0) (Fig. 4.96). With a simple model consideration we discuss the physical relationships. In the inlet cross-section (1) the velocity is constant $=c_\mathrm{m}$. The frictional influence leads to the formation of a boundary layer downstream. The inlet section ℓ ends where this boundary layer (BL) meets the pipe axis (2). From that point on, there is nearly developed flow. In other words: from that point on, boundary layer fills up the whole pipe. It is plausible that the pressure loss in the inlet section is greater than in the case of developed flow. This is because

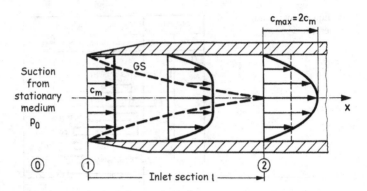

Fig. 4.96 Flow in the inlet section of a circular pipe

1. the wall shear stress is greater and
2. an additional pressure difference is required to change the velocity profile of
   (1) → (2).

We can easily determine the pressure drop in the inlet section. Because of the almost frictionless behaviour of the core flow outside the boundary layer, we calculate on the pipe axis of (1) → (2) with the Bernoulli equation

$$p_0 = p_1 + \frac{\varrho}{2}c_m^2 = p_2 + \frac{\varrho}{2}c_{max}^2 = p_2 + 4\frac{\varrho}{2}c_m^2,$$

$$\Delta p = p_1 - p_2 = 3\frac{\varrho}{2}c_m^2.$$

So there comes the considerable pressure drop of three dynamic pressures. A calculation to determine the length of the inlet section must of course take into account the course of the wall shear stress. With the momentum theorem and energy theorem one can discuss the whole process. An approximate calculation gives

$$\frac{\ell}{D} \approx 0.03 \mathrm{Re}_D, \quad \mathrm{Re}_D = \frac{c_m D}{\nu}. \tag{4.156a}$$

A numerical example illustrates the order of magnitude:

$$\mathrm{Re}_D = 2 \cdot 10^3 \ \left(\text{highest value}\right), \quad \frac{\ell}{D} \approx 60.$$

We hereby calculate the friction pressure drop for the **developed** flow on the same section for comparison:

$$\Delta p = \frac{\varrho}{2}c_m^2 \frac{\ell}{D}\frac{64}{\mathrm{Re}_D} = \frac{\varrho}{2}c_m^2 \cdot 0.03 \cdot 64 = 1.92\frac{\varrho}{2}c_m^2.$$

The additional pressure loss in the inlet section is therefore

$$\Delta p = 1.08\frac{\varrho}{2}c_m^2 = 1.08 p_{dyn}.$$

At this point we give an estimate of the **length of the inlet section**. We use here already a result of the flat plate boundary layer (Fig. 4.97). For the boundary layer thickness δ holds

$$\frac{\delta}{\ell} = \frac{5}{\sqrt{\mathrm{Re}_\ell}} = \frac{5}{\sqrt{\dfrac{U\ell}{\nu}}}. \tag{4.157}$$

We apply this relationship to the inlet section in the pipe. It is clear that we are making a severe simplification here. The spatial influence of the tube is not taken into account. At the end of the inlet section, $\delta = D/2$; U corresponds in the pipe flow to $c_{max} = 2 \cdot c_m$. (4.157) hereby provides

$$\frac{\delta}{\ell} = \frac{D}{2\ell} = \frac{5}{\sqrt{\dfrac{2c_m\ell}{\nu D}D}}, \qquad \frac{\ell}{D} = 0.02\,\mathrm{Re}_D.$$

This agrees, at least as far as the dependence on the Reynolds number is concerned, with (4.156a). The neglected spatial effect and the accelerated core flow (external flow) is expressed in the too small number coefficient.

The pipe inlet flow was studied experimentally by Nikuradse with the following results: The boundary layer has merged along the length considered. The velocity on the pipe axis has the value $c(r = 0) = 1, 9c_m$, at this point, so the profile is even more complete. Another pipe section of the same length follows, in which the velocity on the pipe axis is accelerated $c(r = 0) = 2c_m$ to and the parabolic velocity

Fig. 4.97 Designations for the flat plate boundary layer

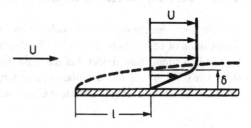

profile is developed. As total length for the inlet section follows from the experiment

$$\frac{\ell_{\mathrm{E}}}{D} = 0.06\,\mathrm{Re}_D, \quad \mathrm{Re}_D = \frac{c_{\mathrm{m}} \cdot D}{\nu}. \tag{4.156b}$$

The additional pressure difference $\Delta p = 1.08 p_{\mathrm{dyn}}$ determined in the model is not far from $\Delta p = 1.16 p_{\mathrm{dyn}}$, the value determined experimentally.

2. **Turbulent flow**

The length of the inlet section is hardly dependent on the Reynolds number. The specifications fluctuate:

$$\frac{\ell}{D} \approx 20\,\mathrm{bis}\,30,$$

depending on how precisely the profile at the end is detected. In principle, this inlet section is shorter than in the laminar case. The turbulent velocity profile developed is almost nearly rectangular and thus already has a close relationship with the inlet profile. The additional pressure drop due to deformation of the velocity profile is also not significant, since only a slight acceleration occurs on the pipe axis. Within the range of validity of the Blasius law, the following applies

$$\overline{p}_1 + \frac{\varrho}{2}\overline{c}_{\mathrm{m}}^2 = \overline{p}_2 + \frac{\varrho}{2}\overline{c}_{\mathrm{m}}^2 = \overline{p}_2 + 1.50\frac{\varrho}{2}\overline{c}_{\mathrm{m}}^2,$$

$$\Delta\overline{p} = \overline{p}_1 - \overline{p}_2 = 0.5\frac{\varrho}{2}\overline{c}_{\mathrm{m}}^2.$$

The number coefficient here is considerably smaller than in the laminar case. It decreases even further with increasing Reynolds number.

4.3.8   Velocity Fluctuations and Apparent Shear Stresses

We now deal with the details of turbulent flows and go back to the Reynolds decomposition of velocities in Sect. 4.3.4. We determine the effect of fluctuation velocities in a simple flow model. Let the main flow direction be in the *x-direction* (Fig. 4.98). We are interested in the impulse force transmitted by the fluctuation velocities on a time-averaged basis to a given control surface.

Fig. 4.98 Flow with
fluctuation velocities

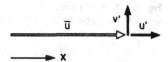

Fig. 4.99 Main flow
direction vertical to the
control surface

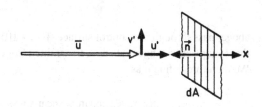

Fig. 4.100 Main flow
direction tangential to the
control surface

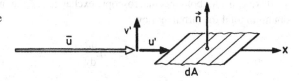

1. Let the control surface be **perpendicular** to the x-axis (Fig. 4.99):

$$\mathrm{d}\boldsymbol{F}_J = -\varrho w \left(\boldsymbol{w}\boldsymbol{n}\right)\mathrm{d}A, \qquad \left|\frac{\mathrm{d}F_{J,x}}{\mathrm{d}A}\right| = \varrho u^2.$$

We form the time average of this normal stress:

$$\overline{\varrho u^2} = \frac{1}{T}\int_0^T \varrho u^2 \mathrm{d}t = \varrho \overline{\left(\overline{u}+u'\right)^2} = \varrho \overline{\left(\overline{u}^2 + 2\overline{u}u' + u'^2\right)} = \varrho \left(\overline{u}^2 + \overline{u'^2}\right). \quad (4.158)$$

Through these fluctuations comes an additional share.

2. The control surface lies **in the** x-direction (Fig. 4.100):

$$\left|\frac{\mathrm{d}F_{J,x}}{\mathrm{d}A}\right| = \varrho u v.$$

The time average of this tangential stress is:

$$\overline{\varrho u v} = \varrho \overline{\left(u+u'\right)v'} = \varrho \overline{u'v'}.$$

We recognize that there is a contribution here only because of two fluctuating
velocities u' and v'. We discuss the sign. We consider particles which, coming from

Fig. 4.101 To determine the sign of the apparent shear stress

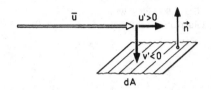

above, pass through the control surface (Fig. 4.101). $u' > 0$, $v' < 0$ lead to $\overline{\tau} > 0$, that is, a positive tangential stress transmitted from the flow to the control surface. We therefore define it as

$$\text{Reynolds apparent shear stress} = \overline{\tau} = -\varrho\overline{u'v'}. \qquad (4.159)$$

If we consider both this macroscopic exchange and the molecular processes, we obtain in total for turbulent flow

$$\overline{\tau}_{\text{ges}} = \eta\frac{d\overline{u}}{dy} - \varrho\overline{u'v'}. \qquad (4.160)$$

According to our derivation, this representation is only valid for a one-dimensional basic flow. In the general case a stress tensor occurs. We will come back to this in the derivation of the Navier-Stokes equations.[24]

We treat two limiting cases of (4.160).

1. **In the immediate vicinity of the wall** ($v' \to 0$), the following occurs

$$\overline{\tau}_{\text{ges}} = \eta\frac{d\overline{u}}{dy}. \qquad (4.161a)$$

This representation in the (laminar) friction sublayer confirms the approaches we made earlier.

2. **In large wall distance**, so-called free turbulence, is $\overline{u} \approx$ constant and thus

$$\overline{\tau}_{\text{ges}} = -\varrho\overline{u'v'}. \qquad (4.161b)$$

We estimate both fractions in the case of pipe flow:

$$|\overline{\tau}_l| = \eta\left|\frac{d\overline{u}}{dy}\right| = \varrho v\frac{\frac{1}{2}\overline{c}_m}{\Delta}, \quad \frac{|\overline{\tau}_l|}{\varrho\overline{c}_m^2} = \frac{1}{2}\frac{v}{\Delta\overline{c}_m} = \frac{1}{2}\frac{1}{\frac{\Delta}{D}\text{Re}_D} = \frac{\lambda_{\text{turb}}}{8}. \qquad (4.162a)$$

---

[24] L. Navier, 1785–1836.

With $\mathrm{Re}D = 10^5$ applies to the smooth pipe $\lambda_{\mathrm{turb}} = 1.7 \cdot 10^{-2}$, so

$$\frac{|\bar{\tau}_1|}{\varrho \bar{c}_{\mathrm{m}}^2} = 2.1 \cdot 10^{-3}.$$

The fluctuation velocities amount to a few percent of the mean velocity; with $\bar{u} \approx \bar{c}_{\mathrm{m}}$ becomes

$$|\overline{\tau_2}| = |\varrho \overline{u'v'}|, \quad \frac{|\overline{\tau_2}|}{\varrho \bar{u}^2} = \left| \frac{u'}{\bar{u}} \frac{v'}{\bar{u}} \right| \approx 5\% \cdot 5\% = 2.5 \cdot 10^{-3}. \tag{4.162b}$$

We see that when the Reynolds number is sufficiently high, the second fraction predominates.

4.3.9 Prandtl's Mixing Length Approach for the Fluctuation Velocities

The problem with the application of (4.160) is that so far we have little information about the fluctuation velocities. We assume here again a mean motion in *x-direction:*

$$u = \bar{u}(y) + u', \quad v = v'$$

and look for a representation of the quantities u', v' by $\bar{u}(y)$. Analogous to the kinetic theory of gases (Sect. 2.3) we introduce the **Prandtl mixing length**. By this we mean the length that a turbulence element travels on average before it mixes with the environment and thus gives up its individuality (Fig. 4.102). This is a macroscopic analogue of the mean free path length of gas kinetics. Specifically, the reasoning goes as follows: A particle from layer y enters the $v' > 0$ at level $y + \ell_1$. There, for the case sketched in Fig. 4.102, it has an under velocity with respect to its surroundings of magnitude

$$\bar{u}(y) - \bar{u}(y + \ell_1) = -\ell_1 \frac{d\bar{u}}{dy}.$$

Prandtl interprets this underspeed as a speed fluctuation in the level $y + \ell_1$, that is,

$$u' = -\ell_1 \frac{d\bar{u}}{dy}. \tag{4.163a}$$

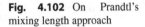
Fig. 4.102 On Prandtl's mixing length approach

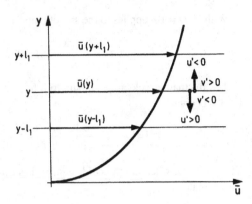

For reasons of continuity, the following is applied accordingly

$$v' = \ell_2 \frac{d\bar{u}}{dy}.$$ (4.163b)

This also gives the correct sign, as is immediately confirmed by Fig. 4.102. For the Reynolds apparent shear stress comes now

$$\bar{\tau} = -\varrho \overline{u'v'} = \varrho \overline{\ell_1 \ell_2} \left(\frac{d\bar{u}}{dy} \right)^2 = \varrho \ell^2 \left(\frac{d\bar{u}}{dy} \right)^2.$$ (4.164a)

For any velocity profile (for example, with $d\bar{u}/dy < 0$), the following applies, taking into account the sign

$$\tau = \varrho \ell^2 \left| \frac{d\bar{u}}{dy} \right| \cdot \frac{d\bar{u}}{dy}.$$ (4.164b)

ℓ is herein a characteristic length measure for mixing in turbulent flows (=Prandtl's mixing length). It must be taken from the experiment as a function of y, therefore this theory is called semi-empirical. It is important to note in (4.164a, b) the dependence on the **square of** the velocity gradient. This indicates typical differences from laminar flow. From (4.160) we get with (4.164a)

$$\bar{\tau}_{\text{ges}} = \eta \frac{d\bar{u}}{dy} + \varrho \ell^2 \left(\frac{d\bar{u}}{dy} \right)^2.$$ (4.165)

We discuss some properties of turbulent flow near a wall.

1. In the (laminar) **friction sublayer** is $\ell \to 0$

$$\overline{\tau}_{\text{ges}} = \eta \frac{d\overline{u}}{dy} = \overline{\tau}_{\text{w}},$$

$$\overline{u}(y) = \frac{\overline{\tau}_{\text{w}}}{\eta} y.$$

With the so-called **wall shear stress velocity**

$$u_{\tau} = \sqrt{\frac{\overline{\tau}_{\text{w}}}{\varrho}} \tag{4.166}$$

becomes

$$\frac{\overline{u}(y)}{u_{\tau}} = \frac{y u_{\tau}}{\nu} = y^{+}. \tag{4.167}$$

The velocity is a linear function of y. y^{+} is a wall distance made suitably dimensionless. The magnitude of the wall shear stress velocity can be estimated from the pipe flow data ($\text{Re}D = 10^{5}, \lambda_{\text{turb}} = 1.7 \cdot 10^{-2}$):

$$\frac{u_{\tau}}{\overline{u}} = \sqrt{\frac{\overline{\tau}_{\text{w}}}{\varrho \overline{u}^{2}}} \approx \sqrt{\frac{\overline{\tau}_{\text{w}}}{\varrho \overline{c}_{\text{m}}^{2}}} = \sqrt{\frac{\lambda_{\text{turb}}}{8}} = 0.05.$$

We thus arrive at the same order of magnitude as the fluctuation velocities and can thus take u_{τ} as a measure of u' and v'.

2. Outside the frictional sublayer, but still near the wall

$$\overline{\tau}_{\text{ges}} = \varrho \ell^{2} \left(\frac{d\overline{u}}{dy} \right)^{2}.$$

We speak here of the **wall turbulence** (Fig. 4.103). Prandtl made the assumption that also here is $\overline{\tau}_{\text{ges}} = \overline{\tau}_{\text{w}} = \text{constant}$ as well as $\ell = \kappa y$ with $\kappa = \text{constant}$.

This will

$$\overline{\tau}_{\text{w}} = \varrho \kappa^{2} y^{2} \left(\frac{d\overline{u}}{dy} \right)^{2}.$$

We can take this as a determination equation for the velocity profile $\overline{u}(y)$. Integration leads with the terms (4.166) and (4.167) to

Fig. 4.103 Velocity profiles in turbulent flow

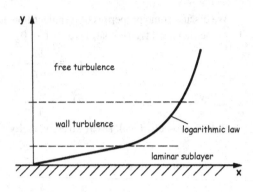

$$\frac{\overline{u}(y)}{u_\tau} = \frac{1}{\kappa}\ln y^+ + C.$$

The two constants κ and C are determined in the experiment. The result is the **logarithmic velocity profile**

$$\frac{\overline{u}(y)}{u_\tau} = 2.5\ln y^+ + 5.5. \tag{4.168}$$

This universal law is valid outside the (laminar) sublayer, so that the singularity at the wall at $y = 0$ is not of importance. At a larger wall distance, the free turbulence follows (4.168). Figure 4.104 contains the two laws (4.167) and (4.168) in semi-logarithmic representation. The estimation of the sublayer thickness Δ carried out in Sect. 4.3.6 leads to the statement:

$$y^+ = \frac{u_\tau y}{v} = \frac{u_\tau \Delta}{v} = \frac{u_\tau}{\overline{u}}\frac{\overline{u}D}{v}\frac{\Delta}{D} \approx 0.05 \cdot 10^4 \cdot 10^{-2} = 5.$$

After a transition area ($5 < y^+ < 30$) the fully turbulent area begins.

4.3.10 General Form of the Navier-Stokes Equations

We proceed analogously to the derivation of Euler's equations of motion. Newton's fundamental law is applied to a mass element (Fig. 4.105). As a result of friction, there is a force acting on each surface element, which we decompose according to the three axial directions. If we relate the respective force to the surface, we obtain

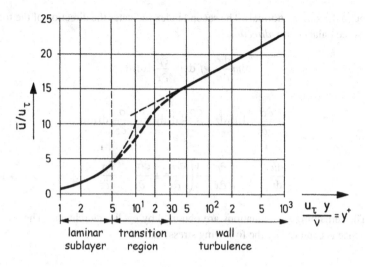

Fig. 4.104 The velocity profiles in semi-logarithmic picture

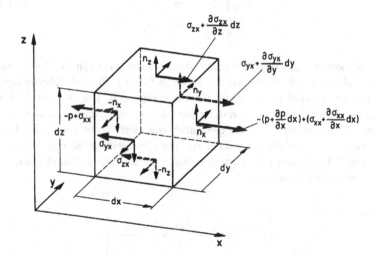

Fig. 4.105 Equilibrium of forces on the mass element with friction

two **shear stresses** in the surface and one **normal stress** perpendicular to the surface. The static pressure p is already split off. The indexing of the stresses is done in such a way that the first index characterizes the position of the surface element (through the surface normal). The second index indicates the direction of the force. The force balance in *x-direction* is

$$dm\frac{du}{dt} = f_x dm - \frac{\partial p}{\partial x}dxdydz$$

$$+ \frac{\partial \sigma_{xx}}{\partial x}dxdydz + \frac{\partial \sigma_{yx}}{\partial y}dxdydz + \frac{\partial \sigma_{zx}}{\partial z}dxdydz,$$

$$\frac{du}{dt} = f_x - \frac{1}{\varrho}\frac{\partial p}{\partial x} + \frac{1}{\varrho}\left(\frac{\partial \sigma_{xx}}{\partial x} + \frac{\partial \sigma_{yx}}{\partial y} + \frac{\partial \sigma_{zx}}{\partial z}\right). \tag{4.169}$$

The remaining two equations are obtained by cyclic interchange. The friction influence is captured by the following stress matrix:

$$\left(\sigma_{ik}\right) = \begin{pmatrix} \sigma_{xx} & \sigma_{yx} & \sigma_{zx} \\ \sigma_{xy} & \sigma_{yy} & \sigma_{zy} \\ \sigma_{xz} & \sigma_{yz} & \sigma_{zz} \end{pmatrix}. \tag{4.170}$$

The moment equilibrium for each surface element states symmetry $\sigma ik = \sigma ki$. Thus, only six independent quantities appear in (4.170). The real difficulty lies in the representation of the σik by the velocity components. That is, it involves the three-dimensional generalization of the (one-dimensional) Newtonian shear stress approach. In continuation of the elementary considerations of Sect. 2.2, a linear relationship between stresses and deformation velocities is postulated for Newtonian fluids. In addition, the following Stokes approach fulfills some obvious, necessary symmetry properties:

$$\left. \begin{aligned} \sigma_{xx} &= 2\eta\frac{\partial u}{\partial x} + \bar{\eta}\left(\frac{\partial u}{\partial x} + \frac{\partial v}{\partial y} + \frac{\partial w}{\partial z}\right), \\ \sigma_{xy} &= \eta(\frac{\partial u}{\partial y} + \frac{\partial v}{\partial x}), \\ \sigma_{xz} &= \eta\left(\frac{\partial u}{\partial z} + \frac{\partial w}{\partial x}\right). \end{aligned} \right\} \tag{4.171}$$

$\bar{\eta}$ is a second coefficient of viscosity in this case. This quantity does not occur in incompressible flow. From the framed part in (4.171) one immediately recognizes the special case of the one-dimensional Newtonian shear stress approach treated earlier. If we restrict ourselves to incompressible flows with $\eta = $ constant, we obtain the Navier-Stokes equations in the form

$$\frac{du}{dt} = f_x - \frac{1}{\varrho}\frac{\partial p}{\partial x} + v\left(\frac{\partial^2 u}{\partial x^2} + \frac{\partial^2 u}{\partial y^2} + \frac{\partial^2 u}{\partial z^2}\right), \quad \dots, \quad \dots, \tag{4.172}$$

where again the two equations not given are obtained by cyclic permutation. In vector form the system is

$$\frac{dw}{dt} = f - \frac{1}{\varrho}\text{grad}\, p + v\,\Delta w. \tag{4.173}$$

In addition, as before, the continuity statement is added, with which four differential equations for $w = (u, v, w)$ and p *are* present. Note that here was assumed $\varrho = $ constant and therefore no further equation is necessary. The **order of** the Navier-Stokes equations is **higher** than that of the Euler differential equations. This allows the condition of adhesion to the surface of the body to be satisfied. The Navier-Stokes equations are **nonlinear** in the same way as the Eulerian equations. This is due to the convective members. Exact solutions are known only in a few cases. We discuss two examples in the next section to learn about properties of these flows.

4.3.11   Special Solutions of the Navier-Stokes Equations

1.   Developed laminar gap flow

We assume a developed layered flow of an incompressible medium in the plane gap ($-h \le y \le +h$, $-\infty < x < \infty$) (Fig. 4.106). This gives $v = w = 0$ and $u = u(y)$. The continuity equation is satisfied. It comes with neglecting the gravity

$$1.\text{Navier   Stokes   Equation}: \frac{1}{\varrho}\frac{\partial p}{\partial x} = v\frac{d^2 u(y)}{dy^2}, \tag{4.174a}$$

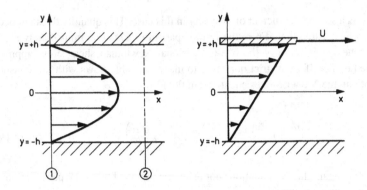

Fig. 4.106   Poiseuille and Couette flow in a plane gap

$$2.\text{and }3.\text{Navier Stokes Equation}: \frac{\partial p}{\partial y} = \frac{\partial p}{\partial z} = 0. \qquad (4.174b)$$

The last two statements give $p = p(x)$. The pressure is constant in the gap transverse to the flow. This is reminiscent of the boundary layer concept. Here, so to speak, the whole gap is filled with boundary layer. (4.174a) leads with $p = p(x)$ immediately to

$$\frac{dp}{dx} = \text{constant}, \quad \frac{d^2u}{dy^2} = \text{constant}. \qquad (4.175)$$

That is, the pressure gradient is constant, and for $u(y)$ comes a very simple second order ordinary differential equation. Integration twice gives

$$u(y) = \frac{1}{\eta}\frac{dp}{dx}\frac{y^2}{2} + Ay + B, \quad A, B = \text{constant}. \qquad (4.176)$$

For the **Poiseuille flow** ($u(\pm h) = 0$), result the parabolic course

$$u(y) = -\frac{h^2}{2\eta}\frac{dp}{dx}\left(1 - \frac{y^2}{h^2}\right) = u_{max}\left(1 - \frac{y^2}{h^2}\right), \qquad (4.177)$$

on the other hand, for the **Couette flow** ($u(-h) = 0$, $u(+h) = U$, $dp/dx = 0$) we obtain the linear function

$$u(y) = \frac{U}{2}\left(1 + \frac{y}{h}\right).$$
(4.178)

The solution for the Poiseuille flow in the gap corresponds completely to the pipe flow discussed earlier. We determine the volume flow as there (b is the width extension of the flow)

$$\dot{V} = u_m 2hb = b\int_{-h}^{h} u\,dy = bu_{max}\int_{-h}^{h}\left(1 - \frac{y^2}{h^2}\right)dy = \frac{2}{3}u_{max}2hb,$$

$$u_m = \frac{2}{3}u_{max}.$$
(4.179)

For the pressure drop along the gap length ℓ thus comes

$$\Delta p = \frac{\rho}{2}u_m^2\frac{\ell}{2h}\frac{24}{Re}, \quad Re = \frac{u_m 2h}{\nu}.$$
(4.180)

The same result is given by the momentum theorem. Let the reader confirm this. The two solutions (4.177) and (4.178) can be linearly superimposed, because in this special case the convective members are omitted and thus the Navier-Stokes equations are linear:

$$u(y) = -\frac{h^2}{2\eta}\frac{dp}{dx}\left(1 - \frac{y^2}{h^2}\right) + \frac{U}{2}\left(1 + \frac{y}{h}\right).$$
(4.181)

The upper edge of the gap is moved with the velocity U, the lower is at rest. The pressure gradient is imposed $dp/dx \lessgtr 0$ on the entire flow field. Depending on the size of dp/dx, interesting velocity distributions results. For example, backflows can occur near the resting wall. Here, the forcing of the upper boundary is not sufficient to overcome the pressure rise throughout the gap (Fig. 4.107). The solution (4.181) describes, the flow in the **lubrication gap** between the shaft and the bearing shell, if the curvature of the gap is neglected (small gap widths). There, the pressure distribution is formed due to the gap geometry, that is, it is built up by the variable distance between shaft and bearing.

2. **Rayleigh-Stokes problem for the plate**

The gap flow just discussed immediately followed on from the pipe flow calculated earlier. Now we discuss a completely different solution of the Navier-Stokes equations, which will lead us directly to the treatment of boundary layer theory.

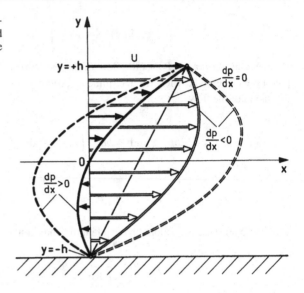

Fig. 4.107 Superposition of Poiseuille and Couette flow in the plane gap with backflow

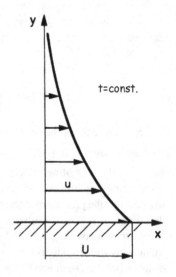

Fig. 4.108 Rayleigh-Stokes flow for the abruptly moving plate

An infinitely extended horizontal plate is abruptly brought to velocity U in a stationary environment. Due to the frictional influence, the fluid above the plate is gradually entrained (Fig. 4.108). The frictional influence spreads further in the transverse direction (y) with time, in other words, the boundary layer grows with time. We now determine its thickness. The flow is developed so that any derivative with respect to x vanishes and $v = 0$. This leaves $u = u(y,t)$. The initial boundary conditions of the problem are

$$t \leq 0: \quad u = 0, \ y \geq 0,$$

$$t > 0: \quad u(0,t) = U, \ u(\infty,t) = 0. \tag{4.182}$$

The Navier-Stokes equations, when eliminating the gravity, provide

$$\frac{\partial u}{\partial t} = v \frac{\partial^2 u}{\partial y^2}, \tag{4.183a}$$

$$\frac{\partial p}{\partial y} = 0. \tag{4.183b}$$

So pressure is constant across to flow direction. This characteristic property of boundary layers is exactly fulfilled here. (4.183a) is of the heat conduction equation type and can be easily solved under the conditions (4.182). We combine y and t into the new, dimensionless variable

$$\frac{y}{\sqrt{vt}} = s$$

together. (4.183a) and (4.182) then go for

$$\frac{u(y,t)}{U} = f(s)$$

into the ordinary differential equation with the following boundary conditions:

$$f''(s) + \frac{s}{2} f'(s) = 0, \quad f(0) = 1, \quad f(\infty) = 0.$$

As a solution we get the error function

$$\frac{u(y,t)}{U} = 1 - \frac{1}{\sqrt{\pi}} \int_0^{y/\sqrt{vt}} \exp\left(-\frac{\xi^2}{4}\right) d\xi = 1 - erf\left(\frac{y}{2\sqrt{vt}}\right). \tag{4.184}$$

This results in a single velocity profile that depends on the variable $y / \left(2\sqrt{vt}\right)$ (Fig. 4.109). From this, the velocity can be determined for any wall distance y at any time t. We want to use (4.184) to determine the thickness δ of that fluid layer

Fig. 4.109 Universal velocity profile for the abrupt moving plate

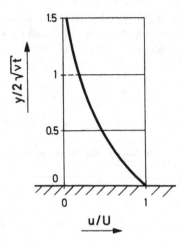

which is entrained as the plate moves. For $y = \delta$ let $u/U = 0, 01$. (4.184) gives for this

$$\delta \approx 4\sqrt{\nu t}. \tag{4.185}$$

This layer thickness therefore increases with time, as is to be expected. For friction and heat conduction processes, the root dependence is characteristic. If one proceeds from the time to the length $\ell = t \cdot U$, then

$$\frac{\delta}{\ell} \approx \frac{4}{\sqrt{\dfrac{U\ell}{\nu}}} = \frac{4}{\sqrt{\mathrm{Re}_\ell}}. \tag{4.186}$$

This inevitably leads to an approach to boundary layer theory, because δ can be understood as the friction layer thickness and the ℓ associated run length. The dependence $\sim 1/\sqrt{\mathrm{Re}_\ell}$ typical for laminar boundary layers occurs, which we will encounter again later. It is important for what follows that the above statements were derived from the Navier-Stokes equations without further neglect.

4.3.12 Introduction to Boundary Layer Theory

For high Re numbers (Re $= U\ell/\nu \gg 1$), friction only plays a role in the boundary layer close to the wall (thickness δ) (Fig. 4.110). There, the increase of the velocity from zero to the value of the external flow occurs. We determine δ for the special case of the longitudinally flowed plate with laminar flow. Moreover, let the flow be

Fig. 4.110 Flow boundary layer on the wing

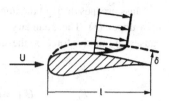

Fig. 4.111 Control surfaces for the plate boundary layer

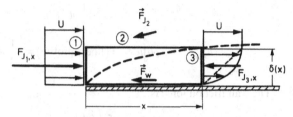

stationary and incompressible. In doing so, we will confirm the relation (4.186). Using the ratios discussed earlier, we immediately see that for the plate for the boundary layer thickness at $x = \ell$ a dependence of the form

$$\frac{\delta}{\ell} = f\left(\mathrm{Re}_\ell\right) \tag{4.187}$$

must exist. With the momentum theorem one can easily determine the function f. For this purpose, we choose a rectangle with side lengths x and $(\delta(x))$ Fig. 4.111) as the control area. The pressure is imposed on the boundary layer by the external flow ($\partial p/\partial y = 0$). In the case of the plate, p is constant in the external space. Therefore, in this case, the pressure is also constant in the boundary layer. We start with a continuity statement. For the mass flows through the control surfaces (1),(2), (3) comes (b = width of the boundary layer considered) $\dot{m}_1 = \varrho U \delta b$,

$$
\left.
\begin{aligned}
\dot{m}_3 &= \varrho b \int_0^\delta u\, dy = \left(\text{especially with the linear profile}\right) = \\
&= \varrho b \frac{U}{\delta} \int_0^\delta y\, dy = \frac{1}{2} \varrho U \delta b, \\
\dot{m}_2 &= \dot{m}_1 - \dot{m}_3 = \varrho b \left(U\delta - \int_0^\delta u\, dy \right) = \left(\text{linear profile}\right) = \\
&= \frac{1}{2} \varrho U \delta b.
\end{aligned}
\right\} \tag{4.188a}
$$

Mass escapes through the upper boundary, since the mass flow in (3) is smaller than upper (1). The boundary layer thus has a displacement effect. This effect causes an impulse force on the surface (2). We get in turn

$$
\left.\begin{array}{ll}
F_{J_1,x} & = \varrho U^2 \delta b, \\[2mm]
F_{J_2,x} & = -\varrho U \int\limits_{(2)} (\boldsymbol{wn})\mathrm{d}A = -U\dot{m}_2 = (\text{linear profile}) = \\[4mm]
& = -\dfrac{1}{2}\varrho U^2 \delta b, \\[4mm]
F_{J_3,x} & = -\varrho b \int\limits_0^\delta u^2 \mathrm{d}y = (\text{linear profile}) = -\dfrac{1}{3}\varrho U^2 \delta b.
\end{array}\right\}
\qquad (4.188\text{b})
$$

The wall friction force is

$$
F_{w,x} = -b\int\limits_0^x \tau_w \mathrm{d}x = -b\int\limits_0^x \eta\left(\frac{\partial u}{\partial y}\right)_w \mathrm{d}x = (\text{linear profile})
$$

$$
= -\eta U b \int\limits_0^x \frac{\mathrm{d}x}{\delta(x)}.
\qquad (4.188\text{c})
$$

The velocity profile in the boundary layer was replaced by a linear function. This serves only to simplify the calculations. The momentum theorem provides the relationship

$$
\frac{U\delta(x)}{6\nu} = \int\limits_0^x \frac{\mathrm{d}x}{\delta(x)}.
$$

Differentiation with respect to x gives the ordinary differential equation for $\delta(x)$

$$
\frac{U}{6\nu}\frac{\mathrm{d}\delta(x)}{\mathrm{d}x} = \frac{1}{\delta(x)},
$$

which can be solved immediately with the initial condition $\delta(0) = 0$. If we replace x by the run length ℓ, we get

$$
\frac{\delta}{\ell} = \frac{3.46}{\sqrt{\dfrac{U\ell}{\nu}}} = \frac{3.46}{\sqrt{\mathrm{Re}_\ell}}.
\qquad (4.189\text{a})
$$

The exact solution of the boundary layer equations gives for the boundary layer thickness $\delta_{1\%}$, at which the external velocity is reached to the value $u(\delta) = 0.99U$, the relation

$$\frac{\delta_{1\%}}{\ell} = \frac{5}{\sqrt{\dfrac{U \cdot \ell}{v}}} = \frac{5}{\sqrt{Re_\ell}}. \qquad (4.189b)$$

We thus confirm the characteristic statements about the boundary layer used earlier. For $Re_\ell \gg 1$ is $\delta/\ell \ll 1$, where the typical dependence occurs $\delta / \ell \sim 1 / \sqrt{Re_\ell}$. A numerical example explains the order of magnitude.

$Re_\ell = 5 \cdot 10^5$ (upper limit), $\delta_{1\%}/\ell \approx 7 \cdot 10^{-3}$, $\ell = 1\,m$, $\delta \approx 7\,mm$.

The wall shear stress $\tau_w(x)$ and the wall friction force W_R are important for the applications.

With the above results comes

$$\tau_w = \eta \left(\frac{\partial u}{\partial y} \right)_w = \left(\text{linear profile} \right) = \eta \frac{U}{\delta(x)} = \frac{\varrho}{2} U^2 \frac{0.557}{\sqrt{\dfrac{Ux}{v}}}. \qquad (4.190)$$

Characteristic is the dependence $\tau_w \sim 1 / \delta(x) \sim 1 / \sqrt{x}$. For the dimensionless coefficient we get

$$c_f = \frac{\tau_w}{\dfrac{\varrho}{2} U^2} = \frac{0.577}{\sqrt{Re_x}}. \qquad (4.191a)$$

The exact solution of the boundary layer equations, that is, without using the linear velocity profile, gives

$$c_{f'} = \frac{0.664}{\sqrt{Re_x}}. \qquad (4.191b)$$

The Reynolds number dependence is preserved, only the number coefficient is changed. By integration we obtain the wall friction force (Fig. 4.112):

$$dW_R = b\tau_w dx, \quad W_R = b\int_0^\ell \tau_w dx.$$

The dimensionless coefficient results in

$$\zeta = \frac{W_R}{\dfrac{\varrho}{2} U^2 b\ell} = \frac{1}{\ell} \int_0^\ell \frac{\tau_w}{\dfrac{\varrho}{2} U^2} dx = \frac{1}{\ell} \int_0^\ell c_f dx,$$

Fig. 4.112 To calculate the
wall friction force

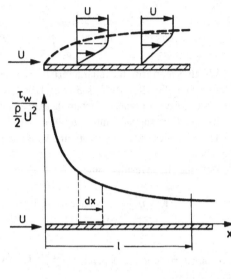

$$\zeta = \frac{1.328}{\sqrt{\mathrm{Re}_\ell}}. \tag{4.192}$$

If the plate is wetted on both sides, the factor 2 comes along. Again, the dependence $\zeta \sim 1/\sqrt{\mathrm{Re}_\ell}$ occurs. If the Reynolds number varies from 10^4 to 10^6, the order of magnitude changes ζ, from 1% to 1‰. These are typical magnitudes of the (laminar) coefficient of friction.

We now come to some basic statements about **laminar boundary layers on curved walls**. The external velocity and the pressure are now no longer constant. They have to be determined by the methods of potential theory (discussed above). At the edge of the boundary layer, we now view them as a known function of x. Depending on this imposed external pressure distribution, different velocity profiles are now established in the boundary layer. Figure 4.113 shows some of these possibilities.

We see that this can lead to significant changes in the near-wall (internal) velocities. The slope of the wall tangent to the velocity profile can become zero (detachment) or even negative (backflow). In contrast, the external velocities change relatively little. On a moderately curved body surface ($u = v = 0$), the first Navier-Stokes equation provides the relation

$$\frac{1}{\varrho}\left(\frac{\partial p}{\partial x}\right)_w = \nu\left(\frac{\partial^2 u}{\partial y^2}\right)_w. \tag{4.193a}$$

Fig. 4.113 Velocity profiles in the boundary layer on a curved surface

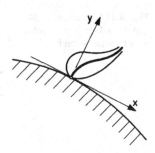

The second Navier-Stokes equation reduces in the case of a high Reynolds number, as in the case of the pipe and gap, to the statement

$$\frac{\partial p}{\partial y} = 0.$$
(4.193b)

If we combine both relations, then with $p = p(x)$ as the boundary layer imprinted pressure distribution of the potential flow on the flowed around body is valid

$$\frac{1}{\varrho}\frac{\mathrm{d}p(x)}{\mathrm{d}x} = \nu\left(\frac{\partial^2 u}{\partial y^2}\right)_{\mathrm{w}}.$$
(4.194)

The right side is a measure of the **curvature of** the velocity profile on the body. The left side we can consider as a given function. At this point the knowledge of the theory of potential flows enters. Figure 4.114 shows a typical case of laminar profile flow (subcritical). The starting point is the pressure distribution on the profile streamline. On the front side of the body we have a complete velocity profile. There is $\mathrm{d}p/\mathrm{d}x < 0$, the flow is accelerated. At the thickness maximum is $\mathrm{d}p/\mathrm{d}x = 0$. According to (4.194), an inflection point in the velocity profile occurs there (P_{w}). At the backside of the body the flow is decelerated, $\mathrm{d}p/\mathrm{d}x > 0$. The **internal** velocities decrease strongly, the wall tangent becomes steeper, the inflection point moves into the interior of the boundary layer. If the wall tangent is normal to the surface (P_{a}), detachment begins. Downstream, backward flow occurs. These retrograde movements can significantly change the potential pressure distribution on the backside of the body.

We distinguish between two typical cases of profile flow:

1. $\mathrm{Re}_\ell = U\ell/\nu < 5 \cdot 10^5$: subcritical flow with laminar separation (Fig. 4.115). Here, laminar flow is present throughout, which comes to separation due to the imposed pressure increase.

Fig. 4.114 Pressure profile
and boundary layer profiles
with laminar airfoil flow

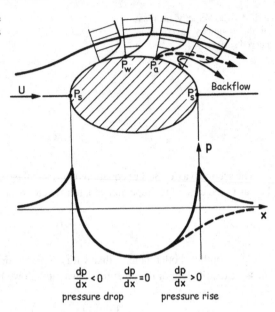

$$\frac{dp}{dx} < 0 \qquad \frac{dp}{dx} = 0 \qquad \frac{dp}{dx} > 0$$

pressure drop   pressure rise

Fig. 4.115 Subcritical airfoil
flow with laminar separation

Fig. 4.116 Supercritical airfoil
flow with turbulent separation

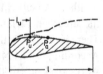

2. $\mathrm{Re}_\ell = U\ell/\nu > 5 \cdot 10^5$: supercritical flow with turbulent separation (Fig. 4.116).
   Here the laminar-turbulent transition (P_u) occurs after the run length ℓ_u. The
   subsequent turbulent boundary layer detaches in P_a.

For the body flowing around it, the critical Reynolds number is

$$\mathrm{Re}_{\mathrm{krit}} = \frac{U\ell_u}{\nu} = 5 \cdot 10^5 \div 10^6, \qquad (4.195)$$

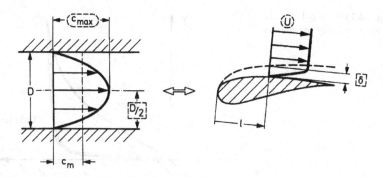

Fig. 4.117 Relationship between Reynolds numbers for pipe and wing

also

$$\frac{\ell_u}{\ell} = \frac{\mathrm{Re_{krit}}}{\mathrm{Re}_\ell} \approx \frac{5 \cdot 10^5}{\mathrm{Re}_\ell}. \tag{4.196}$$

With increasing Reynolds number ℓ_u/ℓ decreases. To $\mathrm{Re}_\ell = 10^7$ for example, applies $\ell_u/\ell = 5 \cdot 10^{-2}$. Here, therefore, a transition already occurs after a short running length.

There is a simple relationship between the critical Reynolds numbers for the pipe and wing, that is, flow through and flow around (Fig. 4.117). There is the correspondence:

$$c_{max} = 2c_m \quad \Rightarrow \quad U,$$

$$\frac{D}{2} \quad \Rightarrow \quad \delta.$$

This allows the Reynolds numbers to be converted as follows:

$$\mathrm{Re}_D = \frac{c_m D}{v} \quad \Rightarrow \quad \frac{U}{2}\frac{2\delta}{v} = \frac{U\delta}{v} = \mathrm{Re}_\delta.$$

The Reynolds number for pipe flow thus corresponds to the Reynolds number formed with the boundary layer thickness δ for the wing.

$$\mathrm{Re}_\delta = \frac{U\delta}{v} = \frac{U\ell}{v}\frac{\delta}{\ell} = \left(\text{Plate boundary layer}\right) = 5\sqrt{\mathrm{Re}_\ell},$$

Fig. 4.118 Turbulent and laminar velocity profile for plate flow

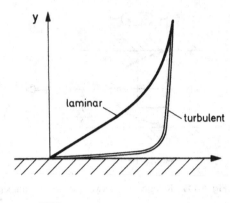

$$\mathrm{Re}_D = 5\sqrt{\mathrm{Re}_\ell}. \tag{4.197}$$

This is the relationship of the Reynolds numbers in the two typical flow problems. The essential difference lies in the different characteristic length scales.

The turbulent velocity profile is always more complete than the laminar one (Fig. 4.118). This statement, known from pipe flow, is equally valid for flow around problems. For the plate we compile the result for the laminar and turbulent boundary layer flow:

$$\frac{W_R}{\frac{\varrho}{2}U^2 b\ell} = \zeta = \begin{cases} \dfrac{1.328}{\sqrt{\mathrm{Re}_\ell}}, & \text{laminar,} \\[2ex] \dfrac{0.074}{\left(\mathrm{Re}_\ell\right)^{1/5}}, & \text{turbulent.} \end{cases} \tag{4.198}$$

The following table explains the orders of magnitude:

	Re_ℓ	ζ
Laminar	10^6 (upper limit)	$1.3 \cdot 10^{-3}$
Turbulent	10^5	$7.4 \cdot 10^{-3}$
	10^6	$4.7 \cdot 10^{-3}$
	$5 \cdot 10^7$	$2.1 \cdot 10^{-3}$

For the same Reynolds number is $\zeta_{\text{turb}} > \zeta_{\text{lam}}$, and the **frictional resistance of** laminar flow is therefore smaller than that of turbulent flow. This fact has led to the development of so-called laminar airfoils in aircraft construction. Here, the transi-

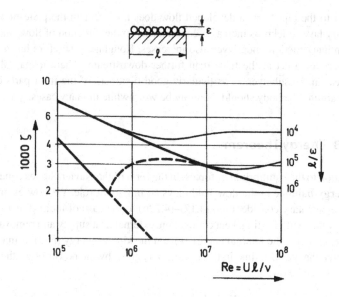

Fig. 4.119 Drag coefficient ζ of the plate as a function of roughness and Reynolds number

tion point is shifted as far as possible to the tail of the body by suitable selection of the airfoil shape, so that the laminar boundary layer is maintained for a long time. It must be noted, however, that in addition to the frictional resistance, **pressure resistance** also occurs. Only both components together result in the **total resistance**. The ratio of the two components can vary within wide limits. For example, only frictional resistance occurs in the case of a plate with the longitudinal flow, whereas only pressure resistance occurs in the case of a transverse plate. We will come back to this in the next section.

For the frictional resistance ζ, of the rough plate, a plot similar to the Nikuradse diagram applies (Fig. 4.119). The plate is hydraulically smooth if the following estimation holds (ε= elevation of the equivalent sand grain roughness):

$$\frac{U\varepsilon}{\nu} = \text{Re}_\ell \frac{\varepsilon}{\ell} \leq 100. \tag{4.199}$$

The allowable roughness ratio ε/ℓ decreases as the Reynolds number increases. For a high-speed aircraft, for example, let $U = 500$m/s, $\nu = 15 \cdot 10^{-6}$m^2/s, $\text{Re}_\ell = 10^8$, $(\varepsilon/\ell)_\text{allowed} = 10^{-6} \varepsilon_\text{allowed} = 3 \cdot 10^{-3}$mm. Figure 4.119 provides $\zeta = 0.002$. If we increase the roughness by a power of ten to $\varepsilon = 3 \cdot 10^{-2}$mm, then $\zeta = 0.0033$. In contrast to the pipe flow, a developed flow does not occur in the plate flow. The boundary layer thickness increases continuously in the direction of flow, and thus the (laminar) friction underlayer also increases. Roughness therefore has a much more serious effect at the front than further downstream, where it may already disappear in the sublayer. A particularly good treatment of the front parts of the flowed around the body should therefore be worthwhile in many cases.

4.3.13 Energy Theorem

In addition to the equilibrium of forces, in the form of the Navier-Stokes equations, the **energy balance** in the flow field plays an important role. We now extend the earlier elementary consideration (4.17)–(4.22b) to the case of multi-dimensional, unsteady flow with friction, heat conduction, external heat supply and removal, etc.

If we combine the internal energy (e) and the kinetic energy ($c^2/2$) as in (4.17), their total change with time in the volume V is given by the power of all the components listed below:

$$\frac{dE}{dt} = \frac{d}{dt} \iiint_v \varrho \left(e + \frac{1}{2}c^2 \right) dV = \tag{4.200}$$

=Power of mass forces f (gravity, electric, magnetic field forces), surface forces (pressure, friction forces), heat flows by conduction, radiation or reactions $\dot{q}_s$. If we use Fig. 4.105 as well as Figs. 4.65 and (4.66), then (4.200) becomes

$$\frac{dE}{dt} = \iiint_v \frac{\partial}{\partial t} \left[\varrho \left(e + \frac{1}{2}c^2 \right) \right] dV + \iint_A \varrho \left(e + \frac{1}{2}c^2 \right)(wn) dA = \left(\text{Gauss's theorem} \right) =$$

$$= \iiint_v \left\{ \frac{\partial}{\partial t} \varrho \left(e + \frac{1}{2}c^2 \right) + div \left[\varrho w \left(e + \frac{1}{2}c^2 \right) \right] \right\} dV$$

$$= \iiint_v \varrho (fw) dV - \iint_A p(wn) dA$$

$$+ \iiint_v \left\{ \frac{\partial}{\partial x} \left(u\sigma_{xx} + v\sigma_{xy} + w\sigma_{xz} \right) + \frac{\partial}{\partial y} \left(v\sigma_{yy} + w\sigma_{yz} + u\sigma_{yx} \right) \right.$$

$$+ \iiint_v \left\{ + \frac{\partial}{\partial z} \left(w\sigma_{zz} + u\sigma_{zx} + v\sigma_{zy} \right) \right\} dV \tag{4.201}$$

$$+ \iint_A \lambda \left(gradT \cdot n \right) dA + \iint_v n\varrho \dot{q}_s dV.$$

Here, again, one can transform all occurring surface integrals into volume integrals with the Gaussian theorem. The integrand must vanish altogether, since the volume V is arbitrary. We obtain from (4.201) the following extensive differential equation

$$\frac{\partial}{\partial t}\left[\rho\left(e+\frac{1}{2}c^2\right)\right]+\frac{\partial}{\partial x}\left[\rho u\left(e+\frac{p}{\rho}+\frac{1}{2}c^2\right)\right]$$

$$+\frac{\partial}{\partial y}\left[\rho v\left(e+\frac{p}{\rho}+\frac{1}{2}c^2\right)\right]+\frac{\partial}{\partial z}\left[\rho w\left(e+\frac{p}{\rho}+\frac{1}{2}c^2\right)\right]$$

$$= \rho\left(uf_x+vf_y+wf_z\right)+u\left(\frac{\partial\sigma_{xx}}{\partial x}+\frac{\partial\sigma_{yx}}{\partial y}+\frac{\partial\sigma_{zx}}{\partial z}\right)$$

$$+v\left(\frac{\partial\sigma_{yy}}{\partial y}+\frac{\partial\sigma_{zy}}{\partial z}+\frac{\partial\sigma_{xy}}{\partial x}\right)+w\left(\frac{\partial\sigma_{zz}}{\partial z}+\frac{\partial\sigma_{xz}}{\partial x}+\frac{\partial\sigma_{yz}}{\partial y}\right)$$

$$+\sigma_{xx}\frac{\partial u}{\partial x}+\sigma_{yy}\frac{\partial v}{\partial y}+\sigma_{zz}\frac{\partial w}{\partial z}+\sigma_{xy}\left(\frac{\partial v}{\partial x}+\frac{\partial u}{\partial y}\right)+\sigma_{yz}\left(\frac{\partial w}{\partial y}+\frac{\partial v}{\partial z}\right)+\sigma_{zx}\left(\frac{\partial u}{\partial z}+\frac{\partial w}{\partial x}\right)$$

$$+\frac{\partial}{\partial x}\left(\lambda\frac{\partial T}{\partial x}\right)+\frac{\partial}{\partial y}\left(\lambda\frac{\partial T}{\partial y}\right)+\frac{\partial}{\partial z}\left(\lambda\frac{\partial T}{\partial z}\right)+\rho\dot{q}_s. \tag{4.202}$$

In this case, the Navier-Stokes eqs. (NS Eq.) (4.169) are used in the dashed portion in the following form: $u \cdot$ (1. NS Eq.) $+ v \cdot$ (2. NS Eq.) $+ w \cdot$ (3. NS Eq.) and the continuity eq. (4.65). If one differentiates everything out and summarizes suitably, then with the abbreviation ϕ for the continuously framed portion (= dissipation) comes

$$\frac{de}{dt}-\frac{p}{\varrho^2}\cdot\frac{d\varrho}{dt}=\frac{dh}{dt}-\frac{1}{\varrho}\cdot\frac{dp}{dt}=T\cdot\frac{ds}{dt}=\frac{1}{\varrho}div\left(\lambda\cdot\mathrm{grad}T\right)+\frac{\phi}{\varrho}+\dot{q}_s. \tag{4.203}$$

Here, on the right-hand side of the equation, there are the heat supplied by heat conduction, friction (dissipation ϕ), radiation as well as by reactions, which contribute to an increase in entropy. This statement corresponds completely to the second law of thermodynamics

$$T\cdot ds = \delta q,$$

in which all the heat supplied to or removed from the mass element is to be absorbed on the right.

From (4.203) with $h = h(p, T)$, $dh = cpdT + 1/\varrho(1 - \beta T)dp$ and the coefficient of thermal expansion $\beta = -1/\varrho(\partial\varrho/\partial T)p$ follows the equation

$$\varrho c_p\frac{dT}{dt} = \lambda\Delta T+\beta\cdot T\frac{dp}{dt}+\phi+\varrho\cdot\dot{q}_s \tag{4.204}$$

which is a generalization of the well-known Fourier heat conduction equation.
We treat **two special cases** with λ = constant and without thermal radiation
$\dot{q}_s = 0$:

1. **Ideal gas**

   with $\beta = 1/T$ follows the heat transport equation

$$\varrho c_p \frac{dT}{dt} = \lambda \Delta T + \frac{dp}{dt} + \phi \qquad (4.205)$$

2. **Incompressible flow**

   with $\beta \to 0$

$$\varrho c_p \frac{dT}{dt} = \lambda \Delta T + \phi \qquad (4.206)$$

For incompressible flow, the isobaric heat capacity plays a decisive role.

The reason for this is that the propagation velocity of heatwaves is much lower
than the propagation velocity of pressure waves, which occurs at the speed of
sound. In the energy equation for incompressible flows, cp must therefore be used
instead of cv [10].

4.3.14 Dissipation and Viscous Potential Flows

The Navier-Stokes equations for incompressible flows with constant material values $\eta = \varrho \cdot \nu$ are given by (4.173):

$$div\, w = 0 \qquad (4.207)$$

$$\frac{dw}{dt} = \frac{\partial w}{\partial t} + w \cdot grad\, w = f - \frac{1}{\varrho} grad\, p + v \cdot \Delta w \qquad (4.208)$$

In addition, there are the transformations (vector identities):

$$\Delta w = grad\left(div\, w\right) - rot\, rot\, w = -rot\, rot\, w \qquad (4.209)$$

$$w \cdot grad\, w = grad\, \frac{w^2}{2} - w \times rot\, w \qquad (4.210)$$

For plane flows, the potential flows with $rot\, w = 0$ are solutions of the Navier-
Stokes equations. The boundary conditions of the potential flows, such as co-mov-
ing edges, are to be fulfilled. The decisive physical processes take place on the
energetic level.

For the dissipation ϕ in the energy equation (4.206) it follows:

$$\phi = \eta \cdot (\mathrm{rot}\, w)^2 + 2 \cdot \eta \cdot \mathrm{div} \left(\mathrm{grad}\, \frac{w^2}{2} - w \times \mathrm{rot}\, w \right) \tag{4.211}$$

For the total dissipation Φ in a flow space V bounded by the area A, it follows:

$$\Phi = \eta \cdot \iiint_V (\mathrm{rot}\, w)^2 \, dV + 2 \cdot \eta \cdot \iint_A \left(\mathrm{grad}\, \frac{w^2}{2} - w \times \mathrm{rot}\, w \right) dA \tag{4.212}$$

The second volume integral has been transformed into a surface integral using Gauss' theorem.

We now consider the special cases:

1. **Dissipation and rotation**

We consider flow fields with constant velocity on the surface (plane Couette flow) or vanishing velocity on the surface (convection in a closed vessel). The dissipation Φ is then given by:

$$\Phi = \eta \cdot \iiint_V (\mathrm{rot}\, w)^2 \, d \tag{4.213}$$

In these cases, the dissipation Φ is given by the rotation alone.

2. **Rotation-free flows with $\mathrm{rot}\, w = 0$ but $\eta \neq 0$**

These flows are viscous potential flows. All dissipation is supplied via the power at the surface:

$$\Phi = 2 \cdot \eta \cdot \iint_A \mathrm{grad}\, \frac{w^2}{2} \, dA \tag{4.214}$$

The realization of viscous potential flows is discussed in Tasks 5.15 and 5.16.

With the stress tensor defined in (4.171), the dissipation ϕ is written as follows:

$$\Phi = \left(\nabla \cdot [\sigma_{ik} \cdot w] \right) - \left(w \cdot [\nabla \cdot \sigma_{ik}] \right) \tag{4.215}$$

This represents the difference of the total power of the frictional forces minus the mechanical power of the frictional forces related to the unit volume.

Another notation follows for Cartesian coordinates by combining the framed component in (4.202) with the stress components (4.171).

$$\phi = \eta \cdot \left[2 \cdot \left(\frac{\partial u}{\partial x} \right)^2 + 2 \cdot \left(\frac{\partial v}{\partial y} \right)^2 + 2 \cdot \left(\frac{\partial w}{\partial z} \right)^2 + \left(\frac{\partial v}{\partial x} + \frac{\partial u}{\partial y} \right)^2 + \left(\frac{\partial w}{\partial y} + \frac{\partial v}{\partial z} \right)^2 + \left(\frac{\partial u}{\partial z} + \frac{\partial w}{\partial x} \right)^2 \right] \tag{4.216}$$

This notation shows the positively defined dissipation. It is zero only where the fluid particles are not deformed.

4.3.15 Resistance and Pressure Drop

The total resistance (1) is the sum of frictional resistance (2) and pressure resistance (3). As far as the measurements are concerned, (1) results from a simple force measurement and (3) by integrating the pressure distribution over the body. The component (2), which is usually more difficult to measure, is then represented as the difference between the terms (1) and (3). The pressure resistance (3) can be of considerable magnitude, since a detachment can substantially change the potential-theoretic pressure distribution near the body's tail. The mutual magnitude relationship of (2) to (3) can be quite different, as we have already noted in the last section. Therefore, an optimization must always take both influences into account. The following statements are valid:

1. **Frictional resistance** should be minimised by ensuring a laminar boundary layer wherever possible.
2. The **pressure resistance** can be reduced by moving the detachment point as far as possible to the rear of the body.

Both influences overlap and partly vary in opposite directions. We will come back to this when discussing the resistance of the sphere.

A Flow around problems

We now return to the basic tasks of fluid mechanics discussed at the very beginning. In this section, we are mainly concerned with the **resistance of** a body in a flow:

$$W = \frac{\varrho}{2} c^2 A c_{\mathrm{w}}. \tag{4.217}$$

ϱ is the density of the flowing medium, c is the incident flow velocity and A is a characteristic reference surface. The dimensionless resistance coefficient cw depends here on all key figures of the problem: Re, M etc.

1. **Key figure-independent body shapes**

Here, a detachment is usually fixed at a sharp edge. A Reynolds number independence exists if the Reynolds number is sufficiently high. The body has primarily pressure resistance.

Flow normal against a **rectangular plate** (Fig. 4.120):

a/b	1	2	4	10	18	∞
c_{w}	1.10	1.15	1.19	1.29	1.40	2.01

Fig. 4.120 Flow against a rectangular plate

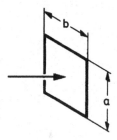

Fig. 4.121 Flow against a circular disc

Fig. 4.122 Flow against a hemisphere

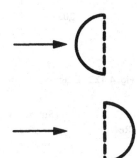

Circular disk (Fig. 4.121) $c_w = 1.11$.
Hemisphere (Fig. 4.122)
Bottomless $c_w = 0.34$,
With bottom $c_w = 0.42$,
Without cover plate $c_w = 1.33$,
With flat cover $c_w = 1.17$.
2. **Key figure-dependent body shapes**
Now the position of the detachment point depends on the Reynolds number. For the **sphere,** for Re < 1 the Stokes formula (= creeping flow) applies $c_w = 24/Re$. In this, 1/3 is pressure drag and 2/3 is frictional drag. For larger Reynolds numbers, (Fig. 4.123):

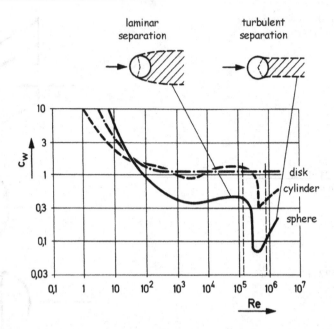

Fig. 4.123 Drag coefficients of sphere, cylinder and disc as a function of Reynolds number

	Subcritical	Supercritical	
Re	$2 \cdot 10^4$ to $3 \cdot 10^5$	$4 \cdot 10^5$	10^6
c_w	0.47	0.09	0.13

The same applies to the **cylinder** (Fig. 4.123):

	Subcritical	Supercritical
Re	$\approx 2 \cdot 10^5$	$5 \cdot 10^5$
c_w	1.2	0.3–0.4

Remarkable in both cases is the rapid drop in resistance during the transition from laminar to turbulent boundary layer flow. When exceeding $Re_{krit} \approx 5 \cdot 10^5$, the pressure resistance decreases considerably, since the turbulent boundary layer detaches later than the laminar one due to the greater energy exchange with the external flow. Prandtl was able to demonstrate this convincingly using a wire that rested on the face of the sphere (tripwire), thereby making the flow turbulent. In the present case, the decrease in pressure resistance overcompensates the increase in frictional resistance in such a way that the total resistance decreases considerably. So here the pressure resistance and its variation play the decisive role. It should be noted that the special shape of the body has a significant influence. If there is a different body, the relation can be reversed. After these preparations, the reader can easily discuss the various cases for himself.

B Flow through problems

The main issue here is the determination of the **pressure loss** Δp_v:

$$\Delta p_v = \frac{\varrho}{2} c_m^2 \zeta_v. \tag{4.218}$$

ζ_v denotes the loss coefficient and, like c_w, depends on the dimensionless characteristics of the problem in question. We have repeatedly encountered such a representation earlier. In the following, we summarize some results for the straight pipe, the diffuser and the manifold. These three examples represent the most important elements of a pipe system.

1. Straight pipe

Here we refer to the explanations given in Sections 4.3.5 and 4.3.6. When the flow is developed

$$\zeta_v = \lambda \frac{\ell}{D}, \tag{4.219}$$

where $\lambda = f(\text{Re}, R/k_s)$ is given by the Nikuradse diagram (Fig. 4.91). If it is the flow in the inlet section, one must go back to (4.218).

2. Diffuser

In this context, we first recall the limiting case of the Carnot diffuser. If we compare (4.218) with (4.106), we find that

$$\zeta_v = \frac{\Delta p_{v,c}}{\frac{\varrho}{2} c_m^2} = \frac{\Delta p_{id} - \Delta p_c}{\frac{\varrho}{2} c_m^2} = \left(1 - \frac{A_1}{A_2} \right)^2. \tag{4.220}$$

The following applies to the diffuser with continuous cross-sectional expansion

$$\zeta_v = k(\alpha) \left(1 - \frac{A_1}{A_2} \right)^2$$

with α as the opening angle:

α	5°	7.5°	10°	15°	20°
k	0.13	0.14	0.16	0.27	0.43

Fig. 4.124 Resistance coefficients of elbows

ζ	1.4	0.76	0.20
Manifold variant			
		1 blade	grid

Fig. 4.125 On the pressure drop in a flowed-through pipe. Boiler (1), long pipe (friction) (2) → (3), Carnot diffuser (3) → (4), short pipe (without friction) (4) → (5)

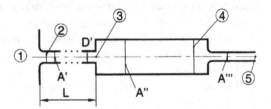

3. Manifold

This element is briefly presented here, as we did not deal with it earlier. As agreed, only the **additional pressure loss** compared to the straight pipe of the same length is given. The order of magnitude can be taken from Fig. 4.124 (Re = 10^5). According to this, already an unprofiled blade reduces the pressure loss coefficient from 1.4 to 0.76. A circular arc grid leads to 0.20. Profiling of the blades reduces the ζ-value again to about 0.10.

Using ζ_{kr} in (4.219), one can define an equivalent pipe length to the manifold loss:

$$\frac{\ell}{D} = \frac{\zeta_{kr}}{\lambda}. \tag{4.221}$$

A manifold with $\zeta_{kr} = 0.20$ thus corresponds, for example, to a straight pipe section (Re = 10^5, $\lambda = 0.02$) of

$$\frac{\ell}{D} = 10.$$

The additive pressure loss of the straight pipe of the same length of a manifold is (Re = 10^5, $\lambda = 0.02$, $\ell/D_h \approx 3$)

$$\zeta = 0.06.$$

This corresponds to half the value of a very good manifold.

We illustrate the results of this section with an example. An incompressible medium flows in a steady state through the pipe given in Fig. 4.125 with the $\dot{V}$ volume flow rate. What we are looking for is the pressure drop $p_1 - p_5$. We use different flow models and the corresponding equations for each section:

$$\text{Bernoulli-Equation}\,(1) \to (2): p_1 = p_2 + \frac{\varrho}{2}c_2^2,$$

$$\text{Pipeflow}\,(2) \to (3): p_2 = p_3 + \frac{\varrho}{2}c_m^2 \frac{L}{D'}\lambda,$$

$$\text{Carnot-Diffusor}\,(3) \to (4): p_4 = p_3 + \frac{\varrho}{2}c_3^2 2\frac{A'}{A''}\left(1 - \frac{A'}{A''}\right),$$

$$\text{Bernoulli-equation}\,(4) \to (5): p_4 + \frac{\varrho}{2}c_4^2 = p_5 + \frac{\varrho}{2}c_5^2,$$

$$\text{Continuity equation}: \dot{V} = c_2 A' = c_m A' = c_3 A' = c_4 A'' = c_5 A'''.$$

By eliminating p_2, p_3 and p_4 one obtains

$$p_1 = p_5 + \frac{\varrho}{2}c_5^2 + \frac{\varrho}{2}c_m^2 \frac{L}{D'}\lambda + \frac{\varrho}{2}c_m^2\left(1 - \frac{A'}{A''}\right)^2,$$

wherein $c_5 = \dot{V}/A'''$ and $c_m = \dot{V}/A'$ are given. According to (4.218), (4.219) and (4.220), the result can be written in the following form:

$$p_1 = p_5 + \frac{\varrho}{2}c_5^2 + \Delta p_{v,\text{Pipe}} + \Delta p_{v,\text{Carnot}}.$$

This is often referred to as the so-called **Bernoulli equation with loss elements**. Thus, on the right-hand side, only the individual pressure losses of the pipe through flow that occurs have to be taken into account additively. This often simplifies the considerations considerably, but requires some experience in dealing with the various loss elements.

4.3.16 Similarity Considerations

By means of examples, we have made a whole series of statements about resistance and pressure loss above. Having made these preparations, we now come to a general discussion of these quantities, particularly as regards their dependence on the ratios introduced earlier (Sect. 4.3.3). This inevitably leads to similarity consider-

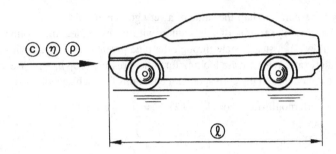

Fig. 4.126 Flow around a body

ations and model laws, which are very important for the applications. Once again, the alternative of flow-around and flow-through problems, already discussed in the introduction, arises. We discuss these problems in turn.

1. **Flow Around Tasks**

We study the flow around a body which a fluid (η, ϱ) with velocity c (Fig. 4.126). The model and the large-scale model should be geometrically similar to each other. Thus, the length ℓ uniquely characterizes the particular body specimen. If we are interested in resistance W, there is a dependence on the form

$$W = f(c, \ell, \varrho, \eta). \tag{4.222}$$

Thus, four independent variables (c, ℓ, ϱ, η) occur here, and accordingly many measurements are required to determine the function f. This characterizes the respective body class. If one goes over to another form of the body, another function takes the place of f. The transition to dimensionless quantities leads to a considerable reduction in the number of variables and thus, of course, also in the number of measurements required. We exemplify this with the above example.

In mechanics, three basic quantities (mass, length, time or force, length, time) occur. Accordingly, we select three from the set of physical quantities (W, c, ℓ, ϱ, η) entering above, for example, (c, ℓ, ϱ) and represent the remaining two dimensionally by power products of these three. We denote the dimension of a quantity a hereby [a]:

$$[W] = [c]^a \cdot [\ell]^b \cdot [\varrho]^c \tag{4.222a}$$

$$[\eta] = [c]^A \cdot [\ell]^B \cdot [\varrho]^C. \tag{4.222b}$$

If we proceed here to force (F), length (L), time (T), then

$$F = L^a T^{-a} L^b F^c L^{-4c} T^{2c},\tag{4.223a}$$

$$FL^{-2}T = L^A T^{-A} L^B F^C L^{-4C} T^{2C}.\tag{4.223b}$$

An exponent comparison for the basic variables leads to

$$a = 2, b = 2, c = 1,$$
$$A = 1, B = 1, C = 1.$$

Thus, the five incoming physical quantities are reduced to the two ratios

$$\frac{W}{\frac{\varrho}{2}c^2\ell^2} = \pi_1, \quad \frac{c\ell}{\nu} = \pi_2.\tag{4.224}$$

The functional relation (4.222) now draws a dependence of the form

$$\frac{W}{\frac{\varrho}{2}c^2\ell^2} = h\left(\frac{c\ell}{\nu}\right) = h(\mathrm{Re}_\ell)\tag{4.225}$$

after it. Thus, the number of required measurements is extraordinarily reduced. The resistance coefficient now depends only on the Reynolds number. In this condensed form, all the earlier representations of resistance can be summarised. (4.225) applies in the same way to model and large-scale. The conversion from one to the other case can be made immediately.

Fig. 4.127 Flow through a horizontal, straight circular pipe

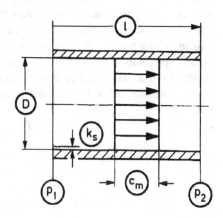

2. Flow through tasks

We start with the horizontal circular pipe (Fig. 4.127). A fluid (v, ϱ) flows through a piece of pipe (ℓ, D, ks) with the mean velocity $\overline{c}_m$. This gives the pressure drop $\Delta \overline{p} = \overline{p}_1 - \overline{p}_2$. There is a dependence of the shape

$$\Delta \overline{p} = \overline{p}_1 - \overline{p}_2 = f\left(\ell, D, k_s, \varrho, v, \overline{c}_m\right). \tag{4.226}$$

We move on to dimensions here as well and represent, for example $\Delta \overline{p}$, D, ks, v by ℓ, ϱ, $\overline{c}_m$. With a similar calculation as above we get this time four parameters

$$\frac{\Delta \overline{p}}{\dfrac{\varrho}{2}\overline{c}_m^2} = \pi_1, \quad \frac{\overline{c}_m D}{v} = \pi_2, \quad \frac{\ell}{D} = \pi_3, \quad \frac{k_s}{D} = \pi_4. \tag{4.227}$$

The dependence (4.226) leads to the functional relationship

$$\frac{\Delta \overline{p}}{\dfrac{\varrho}{2}\overline{c}_m^2} = F\left(\frac{\ell}{D}, \frac{\overline{c}_m D}{v}, \frac{k_s}{D}\right). \tag{4.228}$$

This relationship is generally valid, that is, also in the inlet section. If there is specifically a developed flow, no pipe section is excellent with respect to another. In this case, F must be a linear function of ℓ/D. Thus from (4.228)

$$\frac{\Delta \overline{p}}{\dfrac{\varrho}{2}\overline{c}_m^2} = \frac{\ell}{D}\lambda\left(\mathrm{Re}_D, \frac{k_s}{D}\right), \tag{4.229}$$

with which the earlier representations for laminar and turbulent flows are captured. If one compares (4.226) with (4.229), one immediately recognizes the progress achieved. The number of measurements required is extraordinarily reduced. The Nikuradse diagram provides the function λ, and the Reynolds numbers and the roughness parameter alone enter here as independent variables. Also in this case, (4.229) applies in the same way for model and full-scale model. Geometrically similar flows are described by equal values of ℓ/D and k_s/D. As one considers a nozzle or a diffuser, then an additional parameter occurs, for example, the diameter ratio

$$\frac{D_1}{D_2} = \pi_5 \tag{4.230}$$

At this point, of course, the area ratio A_1/A_2 or a characteristic angle α can be used. If this is taken into account in (4.228), the following applies in general

$$\frac{\Delta \bar{p}}{\frac{\varrho}{2}\bar{c}_m^2} = G\left(\frac{\ell}{D_1}, \frac{D_1}{D_2}, \frac{\bar{c}_m D}{v}, \frac{k_s}{D_1}\right). \tag{4.231}$$

By specialization comes for example, the formula for the Carnot-diffuser (4.220). The reader discusses in detail the necessary conditions for this and compares in particular the earlier derivation with the momentum theorem.

In-Depth Exercises

5

Abstract

In-depth exercises illustrate quasi-stationary and time-dependent flows.

Frictional flows are solved with mechanical and energetic approaches in systems with pipes and connecting elements. Tasks include steady, quasi-steady, and time-dependent flows. The turbulent flow around the flat plate and the Rayleigh-Stokes problem are treated. The compressible flows are analyzed during inflow and outflow and in connection with the Laval nozzle.

The use of wind energy is worked out using a wind turbine and the anemometer. The lift and resistance of bodies flowing around them are dealt with using the example of the parachutist and in aircraft during take-off and cruising flight.

Energetic aspects of viscous potential flows are analyzed using the example of radial and vortex flow in the cylinder gap.

5.1   Task: Flow into a Submersible Tank (Sinking Ship)

The immersion tank shown in Fig. 5.1 fills slowly through the opening in the bottom.

Given:

p_I, ϱ, g, h, A_2, A_3.

Wanted:

$w_2(t)$, replenishment time T.

Solution:

Based on the Bernoulli equation for unsteady flow in the form:

© Springer Fachmedien Wiesbaden GmbH, part of Springer Nature 2022   193
J. Zierep, K. Bühler, *Principles of Fluid Mechanics*,
https://doi.org/10.1007/978-3-658-34812-0_5

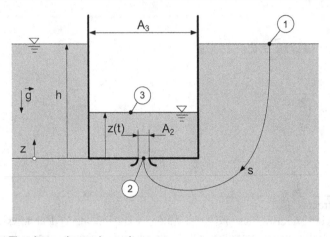

Fig. 5.1 Flow into an immersion tank

$$\int_1^2 \frac{\partial w}{\partial t} ds + \frac{1}{2}\left(w_2^2 - w_1^2\right) + \frac{1}{\varrho}\left(p_2 - p_1\right) + g\left(z_2 - z_1\right) = 0$$

it can be simplified as follows:

For small aspect ratio $A_2 \ll A_3$, the time variation of the velocity along the flow filament s from (1) $\rightarrow$ (2) is small, so that the acceleration term in the Bernoulli equation is negligible. The time dependence is accounted for solely by the time-varying boundary conditions. This flow is called quasi-stationary. From (1) to (2), the Bernoulli equation holds. In (2), the medium flows into the vessel as a free jet. The pressure in the jet is equal to the hydrostatic pressure in the environment at (2) in the vessel: $p_2(t) = p_1 + \varrho g z(t)$. From Bernoulli's equation it now follows for a free surface at rest (1) with $w_1 = 0$ the velocity at the inlet cross-section (2):

$$\frac{\varrho}{2}w_2^2 + p_2 = \frac{\varrho}{2}w_1^2 + p_1 + \varrho g h,$$

$$w_2(t) = \sqrt{2g\left(h - z(t)\right)}.$$

With the continuity equation for the volume flow between (2) and (3)

$$w_2 A_2 dt = A_3 dz.$$

follows the ordinary differential equation, which is solved by separating the variables:

$$dt = \frac{A_3}{A_2} \frac{dz}{w_2(t)} = \frac{A_3}{A_2} \frac{dz}{\sqrt{2g(h - z(t))}}.$$

From the integration, with the initial condition $z = 0$ for $t = 0$:

$$t = \frac{A_3}{A_2} \frac{2h}{\sqrt{2gh}} \left[1 - \sqrt{1 - \frac{z(t)}{h}} \right].$$

For $z = h$ the replenishment time T follows:

$$T = \frac{A_3}{A_2} \frac{2h}{\sqrt{2gh}} = \frac{A_3}{A_2} \sqrt{\frac{2h}{g}}$$

and thus for the temporal change of the liquid level

$$\frac{z(t)}{h} = 1 - \left(1 - \frac{t}{T} \right)^2$$

and for the speed

$$\frac{w_2(t)}{\sqrt{2gh}} = \left(1 - \frac{t}{T} \right).$$

In Fig. 5.2, the time histories of velocity and altitude change are plotted in dimensionless form.

5.2 Task: Oscillating Liquid Column (U-Tube Manometer)

U-tube manometers are often used for pressure measurement. When the liquid column oscillates, there is a transient flow in the U-tube as shown in Fig. 5.3.

Given: ϱ, $L = h_1 + h_2 + \ell$, g.
Wanted: $x(t)$, frequency of oscillation ω.

Fig. 5.2 Time response of
velocity $w(t)$ and liquid level
$z(t)$

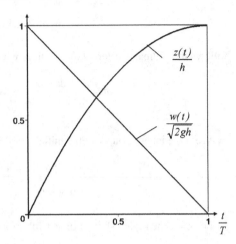

Fig. 5.3 Oscillating liquid
column

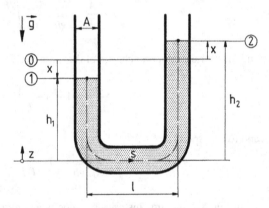

Solution: With constant cross-section A, it follows from the continuity equation
that the velocity $w_1 = w_2 = w(t)$ depends only on the time t, but not on the location
s. The deflection x of the fluid surfaces is equal on both sides. Bernoulli's equation
(4.12a, b), (4.13) then reads for the flow filament s between (1) and (2):

$$\frac{\varrho}{2}w_1^2 + p_1 + \varrho g z_1 = \frac{\varrho}{2}w_2^2 + p_2 + \varrho g z_2 + \varrho \int_1^2 \frac{\partial w}{\partial t}\,\mathrm{d}s.$$

With the pressure equality $p_1 = p_2$ on the two liquid surfaces follows

$$\frac{dw}{dt}\int_1^2 ds + g\left(h_2 - h_1\right) = 0.$$

The length of the flow filament is

$$\int_1^2 ds = L = h_1 + \ell + h_2, \quad \frac{dw}{dt} = \frac{d^2x}{dt^2}, \quad h_2 - h_1 = 2x$$

and the velocity w follows from the change in surface position with time to $w = \dfrac{dx}{dt}$. This gives the following differential equation:

$$\frac{d^2x}{dt^2} + 2g\frac{x}{L} = 0.$$

The solution $x = x_0 \cos(\omega t)$ represents a harmonic oscillation with amplitude x_0 and angular frequency $\omega = \sqrt{\dfrac{2g}{L}}$.

5.3   Task: Time-Dependent Outflow from a Vessel (Start-Up Flow)

From the container with a very large cross-section shown in Fig. 5.4, the liquid flows frictionless through the connected pipe into the environment as soon as the valve at (3) is opened. The temporal development of the outlet velocity $w_3(t)$ up to the steady-state final value $w_{3, \text{st}}$ is of interest.

Fig. 5.4 Frictionless discharge from a vessel with connected pipeline

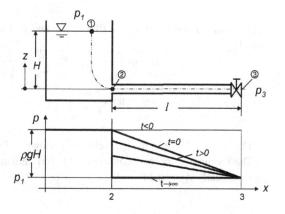

Given: g, H, ℓ.

Wanted: $w_3(t)$, half-life $t \, / \, T$ for the velocity ratio $w_3(t)/w_{3,\,\text{st}}$.

Solution: The problem is solved using the Bernoulli equation for unsteady flow:

$$\frac{\varrho}{2}w_1^2 + p_1 + \varrho g z_1 = \frac{\varrho}{2}w_3^2 + p_3 + \varrho g z_3 + \varrho\int\limits_1^3 \frac{\partial w}{\partial t}\,\mathrm{d}s.$$

With the assumptions: $w_1 = 0$, $p_1 = p_3$, $z_1 = H$ and $z_3 = 0$ follows simplified:

$$\varrho g H = \frac{\varrho}{2}w_3^2 + \varrho\frac{\mathrm{d}w_3}{\mathrm{d}t}\int\limits_2^3 \mathrm{d}s = \frac{\varrho}{2}w_3^2 + \varrho\ell\frac{\mathrm{d}w_3}{\mathrm{d}t}.$$

For the stationary limiting case, the Torricelli formula is obtained:

$$w_{3,\text{st}} = \sqrt{2gH}.$$

With the terminal stationary velocity $w_{3,\,\text{st}}$ follows the nonlinear ordinary differential equation, which can be solved by separating the variables.

$$\frac{\mathrm{d}w_3}{\mathrm{d}t} = \frac{w_{3,\text{st}}^2 - w_3^2}{2\ell},$$

$$\int\frac{\mathrm{d}\dfrac{w_3}{w_{3,\text{st}}}}{1-\left(\dfrac{w_3}{w_{3,\text{st}}}\right)^2} = \frac{w_{3,\text{st}}}{2\ell}\int\mathrm{d}t,$$

with the initial condition $w_3(t = 0) = 0$ follows as a solution with the reference time T:

$$\operatorname{artanh}\frac{w_3}{w_{3,\text{st}}} = \frac{t}{T},\quad T = \frac{2\ell}{w_{3,\text{st}}} = \sqrt{\frac{2}{gH}}\ell,$$

$$\frac{w_3}{w_{3,\text{st}}} = \tanh\frac{t}{T}.$$

The time response of the exit velocity is shown in Fig. 5.5.

The terminal velocity according to the Torricellian efflux formula is reached asymptotically. As half-life follows $\dfrac{t}{T} = 0.55\cdot$

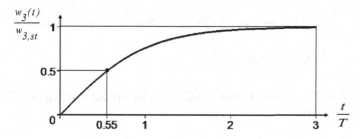

Fig. 5.5 Time response of the velocity

The time-dependent pressure distribution in the pipe is plotted in Fig. 5.4. At time $t = 0$, the pressure along the pipe axis drops linearly from to $p_1 + \varrho g H$ to p_1. The pressure in the pipe is reached constant with p_1 for $t \to \infty$.

5.4 Task: General Outflow Problem

In Fig. 5.6, a liquid flows from the left tank into the right tank via a connected pipe with a shut-off valve. The flow in the pipe is assumed to be developed and the pipe wall hydraulically smooth. The tank surfaces are very large, so the flow can be assumed to be steady. For a given volume flow, the required pressure difference must be determined.

Given: $\varrho, \nu, h, \dot{V}, D, \ell, g, \zeta_E, \zeta_V.$

Wanted: Pressure difference $p_1 - p_6$.

Solution: The solution is given by Bernoulli's equation for the frictionless section and the loss considerations for the frictional section of the system.

Bernoulli's equation (1) $\to$ (2):

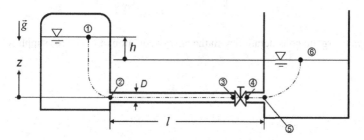

Fig. 5.6 Overflow process between two vessels with connecting pipe line

$$p_1 + \frac{\varrho}{2}w_1^2 + \varrho g z_1 = p_2 + \frac{\varrho}{2}w_2^2 + \varrho g z_2 \text{ with } w_1 = 0 \text{ and } z_2 = 0 \text{ follows}$$

$$p_1 - p_2 = \frac{\varrho}{2}w_2^2 - \varrho g z_1.$$

Frictional flow (2) → (5) with inlet (ζ_E), circular pipe (λ), valve (ζ_V) and Carnot diffuser (free jet):

$$\dot{V} = \frac{\pi D^2}{4}w_2, \quad w_2 = w_3 = w_4, \quad Re = \frac{w_2 D}{\nu} = \frac{4\dot{V}}{\pi D \nu},$$

$$p_2 - p_5 = \frac{\varrho}{2}w_2^2\left(\zeta_E + \frac{\ell}{D}\lambda + \zeta_V\right).$$

Free jet and hydrostatic (5) → (6)

$$p_5 - p_6 = \varrho g z_6.$$

Thus follows for the necessary pressure difference:

$$p_1 - p_6 = \frac{\varrho}{2}w_2^2\left(1 + \zeta_E + \frac{\ell}{D}\lambda + \zeta_V\right) - \varrho g h \quad \text{with } w_2 = \frac{4\dot{V}}{\pi D^2}.$$

Rearranging the question according to the flow velocity $w_2 = w_5$ yields a generalized Torricellian formula:

$$w_2 = \sqrt{\frac{2}{\varrho}\left(\frac{p_1 - p_6 + \varrho g h}{1 + \zeta_E + \frac{\ell}{D}\lambda + \zeta_V}\right)}.$$

An energetic consideration with the energy equation of (1) → (6) supplies:

$$p_1 + \frac{\varrho}{2} w_1^2 + \varrho g z_1 = p_6 + \frac{\varrho}{2} w_6^2 + \varrho g z_6 + \Delta p_V,$$

$$p_1 - p_6 = \varrho g \left(z_6 - z_1 \right) + \Delta p_V = -\varrho g h + \Delta p_V,$$

Pipe inlet : $\Delta p_E = \frac{\varrho}{2} w_2^2 \zeta_E,$

Pipe of length ℓ : $\Delta p_R = \frac{\varrho}{2} w_2^2 \frac{\ell}{D} \lambda,$

Valve : $\Delta p_V = \frac{\varrho}{2} w_2^2 \zeta_V,$

Carnot-Diffusor :

$$\Delta p_C = \frac{\varrho}{2} w_2^2 \zeta_C, \quad \zeta_C = \left(1 - \frac{A_5}{A_6} \right)^2 \rightarrow 1 \quad \text{for } A_6 \rightarrow \infty, \quad \Delta p_C = \frac{\varrho}{2} w_2^2,$$

$$p_1 - p_6 = \frac{\varrho}{2} w_2^2 \left(1 + \zeta_E + \frac{\ell}{D} \lambda + \zeta_V \right) - \varrho g h, \quad w_2 = \frac{4 \dot{V}}{\pi D^2}.$$

5.5 Task: Generalized Overflow Problem

In the flow system shown in Fig. 5.7, consisting of the two vessels connected by a pipeline, the liquid flows through an overpressure at (1) into the vessel with the surface (6). The required pressure difference $p_1 - p_6$ is sought so that the specified volume flow rate $\dot{V}$ occurs. The flow in the pipes is assumed to be developed and the walls are assumed to be hydraulically smooth.

Given: Volume flow rate $\dot{V} = 0.002 \text{m}^3 / \text{s}$, Flow medium water at 20 °C with $\varrho = 998.4 \text{kg/m}^3$, $\nu = 1.012 \cdot 10^{-6} \text{m}^2/\text{s}$, System geometry $h = 7$m, Pipes with $D_1 = 30$mm, $D_2 = 60$mm, $\ell_1 = 50$m, $\ell_2 = 10$m. Loss coefficients in inlet with $\zeta_E = 0.07$ and manifold with $\zeta_K = 0.14$.

Wanted: Pressure difference $p_1 - p_6$.

Solution: Two different solution methods are possible by a mechanical approach based on force balances and an energetic approach along the flow filament coordinate s.

(a) **Mechanical consideration:**

(1) $\rightarrow$ (2) frictionless flow, Bernoulli equation

$$p_1 + \frac{\varrho}{2} w_1^2 + \varrho g z_1 = p_2 + \frac{\varrho}{2} w_2^2 + \varrho g z_2$$

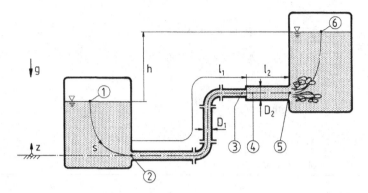

Fig. 5.7 Flow system with pipeline and connecting elements

with the preconditions $w_1 = 0$, $z_2 = 0$, the pressure difference follows

$$p_1 - p_2 = \frac{\varrho}{2} w_2^2 - \varrho g z_1.$$

$(2) \rightarrow (5)$ Frictional pipe flow with loss elements, momentum theorem, continuity, hydrostatics, Reynolds numbers:

$$\mathrm{Re}_1 = \frac{w_2 D_1}{\nu} = 8.39 \cdot 10^4$$

with $w_2 = \dfrac{4}{\pi} \cdot \dfrac{\dot{V}}{D_1^2} = 2.83 \dfrac{\mathrm{m}}{\mathrm{s}}$,

$$\mathrm{Re}_2 = \frac{w_4 D_2}{\nu} = 4.19 \cdot 10^4$$

with $w_4 = w_2 \cdot \dfrac{D_1^2}{D_2^2} = 0.71 \dfrac{\mathrm{m}}{\mathrm{s}}$.

In both pipe sections the flow is turbulent. The pipe resistance coefficients follow from (4.147a) to:

$$\lambda_1 = \frac{0.3164}{\sqrt[4]{\mathrm{Re}_1}} = 0.0186, \quad \lambda_2 = \frac{0.3164}{\sqrt[4]{\mathrm{Re}_2}} = 0.0221.$$

With the pressure loss figures for inlet and manifold and the pressure increase in the Carnot diffuser according to (4.106):

$$P_2 - P_3 = \frac{\varrho}{2} w_2^2 \left(\zeta_E + \frac{\ell_1}{D_1} \lambda_1 + 2\zeta_K \right) + \varrho g z_5,$$

$$P_2 - P_3 = \frac{\varrho}{2} w_2^2 \cdot 31.35 + \varrho g z_5,$$

$$P_3 - P_4 = -\frac{\varrho}{2} w_2^2 \cdot 2 \frac{A_3}{A_4} \left(1 - \frac{A_3}{A_4} \right),$$

$$P_3 - P_4 = -\frac{\varrho}{2} w_2^2 \cdot 0.375,$$

$$P_4 - P_5 = \frac{\varrho}{2} w_4^2 \frac{\ell_2}{D_2} \lambda_2 = \frac{\varrho}{2} w_2^2 \left(\frac{D_1}{D_2} \right)^4 \frac{\ell_2}{D_2} \lambda_2,$$

$$P_4 - P_5 = \frac{\varrho}{2} w_2^2 \cdot 0.230.$$

(5) → (6) Free jet, hydrostatic

$$P_5 - P_6 = \varrho g \left(z_6 - z_5 \right).$$

Summarizing the pressure differences between (1) and (6) results with $z_6 - z_1 = h$:

$$P_1 - P_6 = \frac{\varrho}{2} w_2^2 \left[1 + \zeta_E + \frac{\ell_1}{D_1} \lambda_1 + 2\zeta_K - 2\frac{A_3}{A_4} \left(1 - \frac{A_3}{A_4} \right) + \left(\frac{D_1}{D_2} \right)^4 \frac{\ell_2}{D_2} \lambda_2 \right] + \varrho g h,$$

$$P_1 - P_6 = \frac{\varrho}{2} w_2^2 \cdot 32.19 + \varrho g h = 1.972 \text{bar}.$$

(b) **Energetic consideration:**
Energy equation for steady-state flow system of (1) → (6):

$$P_1 + \frac{\varrho}{2} w_1^2 + \varrho g z_1 = P_6 + \frac{\varrho}{2} w_6^2 + \varrho g z_6 + \Delta p_V.$$

With the assumption of constant mirror height, i.e., $w_1 = 0$, $w_6 = 0$ follows:

$$P_1 - P_6 = \varrho g \left(z_6 - z_1 \right) + \Delta p_V.$$

The pressure losses Δp_V along the coordinate s are composed of:

Pipe inlet$\Delta p_E = \frac{\varrho}{2} w_2^2 \zeta_E$,

Pipe with $\ell_1 \Delta p_{R_1} = \frac{\varrho}{2} w_2^2 \frac{\ell_1}{D_1} \lambda_1$,

Manifold$\Delta p_K = \frac{\varrho}{2} w_2^2 2 \zeta_K$,

Carnot-Diffusor$\Delta p_C = \frac{\varrho}{2} w_2^2 \zeta_1$ with $\zeta_1 = \left(1 - \frac{A_1}{A_2}\right)^2$,

Pipe with $\ell_2 \Delta p_{R_2} = \frac{\varrho}{2} w_4^2 \frac{\ell_2}{D_2} \lambda_2$,

Discharge to container$\Delta p_A = \frac{\varrho}{2} w_4^2 \zeta_A$ with $\zeta_A = 1$,

$$\Delta p_V = \frac{\varrho}{2} w_2^2 \left[\zeta_E + \frac{\ell_1}{D_1} \lambda_1 + 2 \zeta_K + \zeta_1 + \left(\frac{D_1}{D_2}\right)^4 \frac{\ell_2}{D_2} \lambda_2 + \left(\frac{D_1}{D_2}\right)^4 \zeta_A \right],$$

$$\Delta p_V = \frac{\varrho}{2} w_2^2 \cdot 32.19.$$

Thus follows for the pressure difference:

$$p_1 - p_6 = \frac{\varrho}{2} w_2^2 \cdot 32.19 + \varrho g h = 1.972 \text{bar}.$$

5.6   Task: Wind Turbine

The maximum power of a wind turbine is to be determined. Figure 5.8 shows the deceleration of the air through the wind turbine in the flow tube. The course of pressure and velocity along the coordinate x are sketched between the inflow cross-section (1) and the outflow cross-section (5). With the conservation of mass, Bernoulli's equation and the momentum theorem the power can be determined.

Given: Rotor diameter $D = 82 \text{m}$, $w_\infty = 10 \text{m/s}$, $\varrho = 1.205 \text{kg/m}^3$.

Wanted: w_S, holding force F_H, maximum power P_{max}, performance index c_B according to Betz.

Solution: The mass balance for the flow tube enclosing the propeller as a control space provides the mass flow $\dot{m}$:

Fig. 5.8 Principle
of a wind turbine
with an outer control
space and a control
space around the
rotor as well as the
pressure and velocity
distribution in the
direction of flow

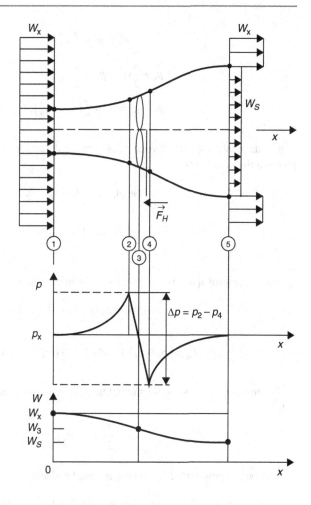

$$\varrho w_\infty A_1 = \varrho w_3 A_3 = \varrho w_S A_5 = \dot{m}.$$

Between cross sections (1) and (2), as well as (4) and (5), Bernoulli's equation
(4.36) is valid. With the prerequisite $A_2 \approx A_3 \approx A_4$ follows $w_2 \approx w_3 \approx w_4$ and thus
the pressure difference:

$$p_\infty + \frac{\varrho}{2} w_\infty^2 = p_2 + \frac{\varrho}{2} w_3^2,$$

$$p_4 + \frac{\varrho}{2} w_3^2 = p_\infty + \frac{\varrho}{2} w_S^2,$$

$$p_2 - p_4 = \Delta p = \frac{\varrho}{2} \left(w_\infty^2 - w_S^2 \right).$$

For the control space between cross-sections A_1 and A_5 it follows with the momentum theorem (4.99)

$$\varrho w_\infty^2 A_1 - \varrho w_S^2 A_5 - F_H = 0,$$

$$F_H = \varrho w_\infty^2 A_1 - \varrho w_S^2 A_5.$$

For the control space between (2) and (4) around the rotor follows:

$$p_2 A_3 - p_4 A_3 - F_H = 0,$$

$$F_H = \left(p_2 - p_4 \right) A_3 = \Delta p A_3 = \frac{\varrho}{2} \left(w_\infty^2 - w_S^2 \right) A_3.$$

By equating the results for the holding force, the velocity in cross section A_3 follows

$$w_3 = \frac{1}{2} \left(w_\infty + w_S \right).$$

The output power P of the system is calculated as

$$P = F_H w_3 = \frac{\varrho}{4} A_3 \left(w_\infty^2 - w_S^2 \right) \left(w_\infty + w_S \right) = \frac{\varrho}{4} A_3 \left(w_\infty^3 + w_\infty^2 w_S - w_S^2 w_\infty - w_S^3 \right).$$

We obtain the maximum value from the extreme value observation:

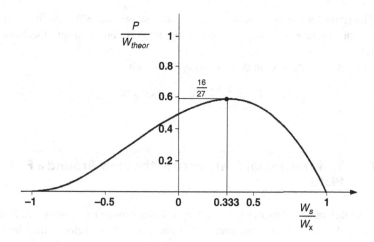

Fig. 5.9 Dependence of the power P/P_{theor} on the speed deceleration w_S/w_∞

$$\frac{dP}{dw_S} = \frac{\varrho}{4} A_3 \left(w_\infty^2 - 2w_\infty w_S - 3w_S^2 \right) = 0, \quad w_{S_{1,2}} = \begin{cases} \dfrac{1}{3} w_\infty & \text{max.,} \\ -w_\infty & \text{min.,} \end{cases}$$

$$\frac{d^2 P}{dw_S^2} = \frac{\varrho}{4} A_3 \left(-2w_\infty - 6w_S \right) < 0 \quad \rightarrow \text{Maximum}$$

at the rate of $w_S = w_\infty/3$ follows

$$P_{\max} = \frac{16}{27} \cdot \frac{\varrho}{2} A_3 w_\infty^3 = \frac{8}{9} \cdot \frac{\varrho}{2} A_1 w_\infty^3.$$

In relation to the theoretically possible energy flow $P_{\text{theor}} = \dfrac{1}{2} \varrho A_3 w_\infty^3$ through the propeller cross-section, the coefficient of performance (Betz number) follows (Fig. 5.9):

$$c_{\text{B}} = \frac{P_{\max}}{\dfrac{1}{2} \varrho A_3 w_\infty^3} = \frac{16}{27} = 0.593 = \text{Key performance} \left(\text{Betz number} \right).$$

The result for the maximum power P_{max} also shows that $8/9 \approx 88.9\%$ kinetic energy of the mass flow passing through the rotor is converted into mechanical power.

The output of the system is $w_\infty = 10\,\text{m/s} = 36\,\text{km/h}$

$$P_{max} = \frac{\varrho}{2} \frac{\pi D^2}{4} w_\infty^3 c_B = 1887\,\text{kW}.$$

5.7   Task: Frictional Resistance in the Flow Around a Flat Plate

A plane thin plate of length ℓ and width b is flowed around by a viscous medium. The boundary layer flow is assumed to be turbulent, so that a velocity distribution shown in Fig. 4.118 is the basis. The wall friction force F_w is to be calculated using the momentum theorem according to Fig. 4.111.

Given:
Boundary layer thickness $\dfrac{\delta}{\ell} = \dfrac{0.37}{\text{Re}^{\frac{1}{5}}}$, velocity distribution $\dfrac{u(y)}{U} = \left(\dfrac{y}{\delta}\right)^{\frac{1}{7}}$.

Wanted: Wall friction force $F_{w,\,x}$.

Solution: The solution is done with a balance of forces in the *x-direction* (momentum theorem):

Continuity:

$$\dot{m}_1 = \varrho U b \delta, \quad \dot{m}_3 = \varrho b \int_0^\delta u\,\mathrm{d}y,$$

$$\dot{m}_2 = \dot{m}_1 - \dot{m}_3 = \int_0^\delta \varrho b U \left(1 - \frac{u}{U}\right)\mathrm{d}y = \varrho b U \delta^*,$$

$$\delta^* = \int_0^\delta \left(1 - \frac{u}{U}\right)\mathrm{d}y = \text{Displacement thickness.}$$

Impulse forces:

$$F_{J_1,\,x} = \varrho U^2 b \delta, \quad F_{J_2,\,x} = -\varrho U \int_{(2)} (\boldsymbol{w}\boldsymbol{n})\,\mathrm{d}A = -U\dot{m}_2 = -\varrho U^2 b \delta^*,$$

$$F_{J_3,\,x} = -\varrho b \int_0^\delta u^2\,\mathrm{d}y.$$

Wall friction force: $F_{w,x} = -b\int_0^x \tau_w dx.$

Momentum theorem:

$$\varrho U^2 b\delta - \varrho U^2 b\delta^* - \varrho b\int_0^\delta u^2 dy - b\int_0^x \tau_w dx = 0$$

$$\varrho U^2 \delta - \varrho U^2 \delta^* - \varrho \int_0^\delta u^2 dy - \int_0^x \tau_w dx = 0$$

$$\varrho U^2 \int_0^\delta \frac{u}{U}\left(1-\frac{u}{U}\right)dy - \int_0^x \tau_w dx = 0$$

$$\frac{F_{w,x}}{\dfrac{\varrho}{2}U^2 b\ell} = \int_0^\delta \left(\frac{y}{\delta}\right)^{\frac{1}{7}}\left[1-\left(\frac{y}{\delta}\right)^{\frac{1}{7}}\right]dy = \frac{14}{72}\delta = \frac{0.074}{Re_1^{1/5}}.$$

For the overflow area on the top of the plate, the following is due $c_F = \dfrac{0.074}{Re^{1/5}}.$

If the plate is flowed around on both sides, the value $c_W = 2c_F$ follows for the resistance coefficient.

5.8   Task: Suddenly Accelerated Plate (Rayleigh-Stokes Problem)

For the Rayleigh-Stokes problem of the suddenly started plate (Sect. 4.3.11), determine the energy balance using the control space in Fig. 5.10 with the lengths in the ℓ_x in the x-direction, ℓ_y in the y-direction, and $\ell_z = 1$ the depth unit.

Given: Velocity distribution (4.185) $\dfrac{u(y,t)}{U} = 1 - erf\left(\dfrac{y}{2\sqrt{vt}}\right).$

Wanted: Power of the wall shear stress L, the dissipation Φ and the time change of the kinetic energy dE_{kin}/dt.

Fig. 5.10 Control space with designations

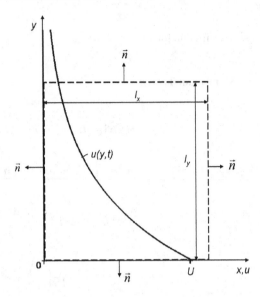

Solution: The energy balance (power balance) for the control volume V with the surface A and the velocity vector w is: $\dfrac{\mathrm{d}}{\mathrm{d}t}E_{\mathrm{kin}}+L+\Phi=0$. The individual proportions result in

$$\frac{\mathrm{d}}{\mathrm{d}t}E_{\mathrm{kin}}=\frac{\mathrm{d}}{\mathrm{d}t}\int_{V}\frac{\varrho}{2}w^{2}\mathrm{d}V=\int_{V}\frac{\partial}{\partial t}\left(\frac{\varrho}{2}w^{2}\right)\mathrm{d}V+\int_{A}\frac{\varrho}{2}w^{2}\left(w\cdot n\right)\mathrm{d}A$$

$$=\ell_{x}\int_{0}^{\ell_{y}}\varrho\cdot u\frac{\partial u}{\partial t}\mathrm{d}y=\ell_{x}\int_{0}^{\ell_{y}}u\frac{\partial\tau}{\partial y}\mathrm{d}y$$

$$=\ell_{x}\left\{u(y,t)\cdot\tau(y,t)\big|_{y=0}^{\ell_{y}}-\int_{0}^{\ell_{y}}\tau\frac{\partial u}{\partial y}\mathrm{d}y\right\}.$$

This takes into account the momentum theorem in the form $\varrho\dfrac{\partial u}{\partial t}=\dfrac{\partial\tau}{\partial y}$. The limiting process $\ell_{y}\to\infty$ leads to the final result:

$$\frac{d}{dt}E_{kin} = -\ell_x U\tau(0,t) - \ell_x \int_0^\infty \tau \frac{\partial u}{\partial y}dy = -L - \Phi,$$

$$L = \ell_x U\tau(0,t) = \frac{-\eta \ell_x U^2}{\sqrt{\pi v t}} = \text{Power component of the wall shear stress,}$$

$$\Phi = \ell_x \int_0^\infty \tau \frac{\partial u}{\partial y}dy = \frac{\eta \ell_x U^2}{\sqrt{2\pi v t}} = \text{Dissipation,}$$

$$\frac{d}{dt}E_{kin} = \frac{\eta \ell_x U^2}{\sqrt{\pi v t}}\left(1 - \frac{1}{\sqrt{2}}\right).$$

This is the solution for the suddenly accelerated plate. The energy transferred from the plate to the medium per unit time (power of the wall shear stress) is found in the increase of kinetic energy and dissipation of the medium.

Corresponding results are obtained for the suddenly delayed and for the periodically oscillating plate. See K. Bühler, J. Zierep: Energetic considerations of the Rayleigh-Stokes problem. Proc. Appl. Math. Mech. PAMM **5**, 539–540 (2005).

5.9   Task: Compressible Inflow and Outflow

Gas dynamic test facilities can be realized with a vacuum vessel with connected flow channel. For a certain time t_{krit}, supersonic velocity can be reached in the measuring section. Determine the suction time of a supersonic channel using dimensional analysis (Fig. 5.11).

Given: p_1, ϱ_1, p_{30}, V, A^*, $\kappa = \dfrac{c_p}{c_v}$.

Searched for: t_{krit}= critical suction time for supersonic velocity in the measuring section.

Solution: We assume the following relationship between the physical dimensional quantities:

$$t_{krit} = f\left(p_1, \varrho_1, p_{30}, V, A^*, \kappa = \frac{c_p}{c_v}\right).$$

Fig. 5.11 Vacuum tank with connected measuring section

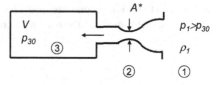

In the context of mechanics, this results in three other dimensionless ratios in addition to the ratio of the specific heats κ:

$$\pi_1 = \frac{t_{\text{krit}} \cdot A^* \cdot a_1}{V}, \quad \pi_2 = \frac{p_1}{p_{30}}, \quad \pi_3 = \frac{V^2}{A^{*3}}, \quad \pi_4 = \kappa.$$

with the speed of sound (4.51) $a_1 = \sqrt{\kappa p_1 / \varrho_1}$ and the dependence $\pi_1 = F(\pi_2, \pi_3, \pi_4)$. This gives the following relationship:

$$t_{\text{krit}} = \frac{V}{A^* \cdot a_1} \cdot H\left(\frac{p_1}{p_{30}}, \frac{V^2}{A^{*3}}, \kappa\right).$$

There $t_{\text{krit}} \sim \dfrac{1}{A^*}$ must be, comes simplifying

$$t_{\text{krit}} = \frac{V}{A^* \cdot a_1} \cdot h\left(\frac{p_1}{p_{30}}, \kappa\right).$$

The mass balance $\dfrac{d}{dt}(p_3 \cdot V) = -\varrho^* \cdot a^* \cdot A^*$ gives for the quasi-stationary flow with the other basic equations for the critical suction time

$$t_{\text{krit}} = \frac{V}{A^* \cdot a_1} \cdot \frac{\left(\dfrac{2}{\kappa+1}\right)^{\frac{\kappa}{\kappa-1}} - \dfrac{p_{30}}{p_1}}{\kappa \cdot \left(\dfrac{2}{\kappa+1}\right)^{\frac{\kappa+1}{2(\kappa-1)}}}.$$

The two borderline cases are important

$$\frac{p_{30}}{p_1} = \begin{cases} \left(\dfrac{2}{\kappa+1}\right)^{\frac{\kappa}{\kappa-1}} = 0.528, \\ 0, \end{cases} \quad t_{\text{krit}} = \begin{cases} 0 \\ 0.652 \cdot \dfrac{V}{A^* \cdot a_1}. \end{cases}$$

The first case represents the critical pressure ratio (4.62) and provides a vanishing suction time. The second case provides the maximum suction time in which the speed of sound prevails in the narrowest cross-section:

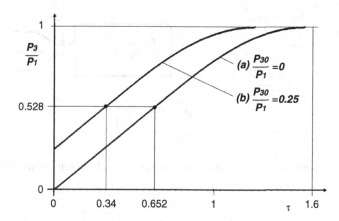

Fig. 5.12 Relationship between the dimensionless pressure ratio p_3/p_1 and the dimensionless suction time τ

$$t_{\text{krit}} = \frac{V}{A^* \cdot a_1} \cdot \frac{1}{\kappa} \cdot \sqrt{\frac{2}{\kappa+1}} = 0.652 \cdot \frac{V}{A^* \cdot a_1}.$$

Figure 5.12 shows this result in dimensionless form $\tau = t_{\text{krit}} A^* a_1 / V$. The critical suction time τ_{krit} decreases with increasing initial pressure ratio p_{30}/p_1.

Numerical example of the system in the Institute of Fluid Mechanics at the Karlsruhe Institute of Technology KIT:

$$a_1 = 330\frac{\text{m}}{\text{s}}, \quad V = 30\text{m}^3, \quad A^* = 30\text{cm}^2,$$

$$t_{\text{krit}} = \frac{V}{A^* \cdot a_1} \cdot \tau_{\text{krit}}, \quad t_{\text{krit}} = \frac{30\text{m}^3 \cdot 0.652}{30 \cdot 10^{-4}\text{m}^2 \cdot 330\frac{\text{m}}{\text{s}}} \approx 20\text{s}.$$

Reversing the motion (Fig. 5.13), the question of the blow-down time of a blow-down channel arises (Abboud, Bühler).

In this case, the dimensional analysis provides the correlation for the critical blowing time:

Fig. 5.13 Discharge from a
vessel under over pressure
into the atmosphere

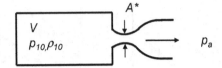

$$t_{\text{krit}} = \frac{V}{A^* \cdot a_{10}} \cdot F\left(\frac{p_{10}}{p_a}, \kappa\right).$$

The conservation laws for mass and energy gives the result for quasi-steady flow:

$$t_{\text{krit}} = \frac{V}{A^* \cdot a_{10}} \cdot \frac{2}{\kappa - 1} \cdot \frac{\left(\dfrac{p_{10}}{p_a}\right)^{\frac{\kappa-1}{2\kappa}} - \left(\dfrac{\kappa+1}{2}\right)^{\frac{1}{2}}}{\left(\dfrac{2}{\kappa+1}\right)^{\frac{1}{\kappa-1}}},$$

$$t_{\text{krit}} \to 0, \quad \frac{p_{10}}{p_a} \to \left(\frac{\kappa+1}{2}\right)^{\frac{\kappa}{\kappa-1}} = 1.894.$$

This corresponds to the critical pressure ratio $\dfrac{p^*}{p_{10}} = 0.528.$.

The critical blowing time t_{krit} increases with increasing ratio of resting pressure to ambient p_{10}/p_a pressure.

Numerical example:

$$a_1 = 330\frac{\text{m}}{\text{s}}, \quad V = 30\text{m}^3, \quad A^* = 30\text{cm}^2, \quad \frac{p_{10}}{p_a} = 5,$$

$$t_{\text{krit}} = \frac{V}{A^* \cdot a_1} \cdot \tau_{\text{krit}}, \quad t_{\text{krit}} = \frac{30\text{m}^3 \cdot 1.286}{30 \cdot 10^{-4}\text{m}^2 \cdot 330\dfrac{\text{m}}{\text{s}}} \approx 39\text{s}.$$

In the case of blowing out with $p_{10}/p_a > 3, 17$, higher blowing times can be achieved compared to flowing into the vacuum from the atmosphere.

For further questions see:

M. Abboud:	The quasi-steady flow of gas during the emptying and filling of a container, Fluid Mechanics and Fluid Machinery **33**, 59–70 (1983).
K. Buehler:	Gas dynamics study of the Joule overflow experiment, Heat and Mass Transfer **23**, 27–33 (1988).

5.10   Task: Laval Nozzle Flow

Using the differential equation (4.55) and the relation (4.57), discuss the flow in a Laval nozzle with two narrowest cross sections (A_1, A_3) and one widest cross section (A_2) as shown in Figs. 5.14 and 5.15. When is a compression shock possible at all?

Given: Nozzle cross sections A_1, A_2, A_3, Shock Mach number $M_s = 2$.

Searched: Mach number distribution in the nozzle, location of the shock.

Solution: If $A_1 = A_3$ (Fig. 5.14), there are simultaneous critical conditions in cross sections 1 and 3. There they are saddle points of the integral curves (4.55), while 2 is a vortex point. This follows from (4.57) for the two singular points.

A normal shock between 1 and 3 is not possible. This would otherwise lead to a decrease in the rest variables and the critical values and thus reduce the mass flow. A_3 is too low to fulfil the continuity (blocking!).

If $A_1 < A_3$ (Fig. 5.15), we have a model for a supersonic duct with sound passage in 1. A shock between 1 and 3 is possible if the adjustable diffuser in 3 is opened by as much as the decrease in the rest magnitudes dictates.

The plot for the resting pressure decrease at normal shock (Fig. 4.39) gives:

$$\frac{A_1}{A_3} = \frac{\hat{\varrho}^* \cdot \hat{a}^*}{\varrho^* \cdot a^*} = \frac{\hat{\varrho}_0}{\varrho_0} = \frac{\hat{p}_0}{p_0} = f\left(M_s\right)$$

$$f\left(M_s\right) = \left[1 + 2\frac{\kappa}{\kappa+1}\left(M_s^2 - 1\right)\right]^{\frac{-1}{\kappa-1}} \cdot \left[1 - \frac{2}{\kappa+1}\left(1 - \frac{1}{M_s^2}\right)\right]^{\frac{-\kappa}{\kappa-1}}.$$

For $M_s = 2$, the area ratio gives $\dfrac{A_1}{A_3} = 0.721$.

Fig. 5.14 Laval nozzle with two narrowest cross sections $A_1 = A_3$

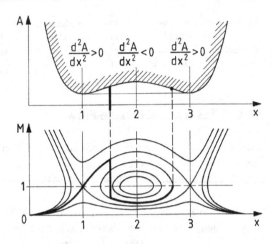

Fig. 5.15 Laval nozzle with two cross sections $A_1 < A_3$

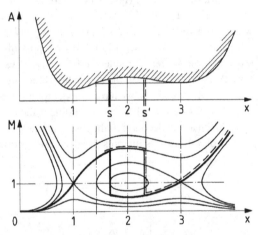

5.11 Task: Speed During Free Fall (Parachutist)

Figure 5.16 shows the parachutist with the acting forces. In a stationary case, the gravitational force and the flow resistance are in equilibrium. The speed of fall w is required.

Given: $D = 8$ m, $m = 90$ kg, $\varrho_L = 1.188\dfrac{\text{kg}}{\text{m}^3}$, $c_W = 1.33$.

Wanted: Stationary fall velocity w.

Fig. 5.16 Skydiver

Solution: If the umbrella shape corresponds to an open hemisphere, then follows (Fig. 4.122) the drag coefficient $c_W = 1.33$. With the force equilibrium of gravity and drag force, neglecting the buoyancy, it follows

$$F_W = mg = \frac{\varrho}{2} w^2 A c_W.$$

From this expression the fall velocity is given by

$$w = \left(\frac{8mg}{\pi D^2 \varrho c_W} \right)^{\frac{1}{2}} \approx \left(\frac{8 \cdot 90\,\text{kg} \cdot 9.81 \frac{\text{m}}{\text{s}^2}}{\pi \cdot 8^2 \text{m}^2 \cdot 1.188 \frac{\text{kg}}{\text{m}^3} \cdot 1.33} \right)^{\frac{1}{2}} \approx 4.7 \frac{\text{m}}{\text{s}} \approx 17 \frac{\text{km}}{\text{h}}.$$

In reality, the drag coefficient c_W is lower due to the porosity of the screen and the velocity is thus somewhat higher.

5.12 Task: Lift Coefficients of Aircraft (Take-Off and Cruise)

The Airbus A380 aircraft has a take-off weight of $m = 560,000$ kg. The wing cross-section is given by the reference area A. The engines have a thrust of $F_S = 1244$ kN. The speed at takeoff is w_S and increases to w at cruise altitude of

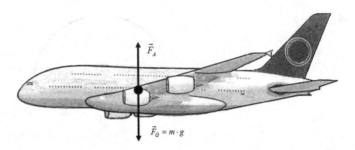

Fig. 5.17 Aircraft in cruise flight

11 km. The lift coefficients for cruise c_A and for take-off $c_{A, max}$ are to be determined (Fig. 5.17).

   Given: $A = 845$ m², $m = 560,000$ kg, $g = 9.81$ m/s², $F_S = 1244$ kN.

z in m	p in Pa	ϑ in °C	ϱ in kg/m³	a in m/s	w in m/s	w in km/h	M
0	101.325	15	1.225	340.26	69.44	250	0.204
11.000	22.614	−56.5	0.364	295.04	262.58	945	0.890

   Wanted: Lift coefficients c_A (cruise) and $c_{A, max}$ (takeoff).

   Solution: For the lift coefficient c_A at cruising altitude $z = 11$ km follows with $F_A = F_G = mg$:

$$c_A = \frac{F_A}{\frac{\varrho}{2} w^2 A} = \frac{5.6 \cdot 10^5 \, \text{kg} \cdot 9.81 \frac{\text{m}}{\text{s}^2}}{\frac{0.364 \frac{\text{kg}}{\text{m}^3}}{2} \cdot \left(262.58 \frac{\text{m}}{\text{s}}\right)^2 \cdot 845 \text{m}^2} = 0.52.$$

   At takeoff, the lift coefficient is maximum $c_{A, max}$ with the data on the ground at $z = 0$. Due to the aircraft attitude with the angle of attack α, the thrust force of the engines F_S contributes by the vertical thrust component with a fraction of $F_S \cdot \sin \alpha$, so that $F_A = F_G - F_S \cdot \sin \alpha$ results for the lift force. For an angle of attack of $\alpha = 10°$ follows

$$c_{A, max} = \frac{F_A}{\frac{\varrho}{2} w^2 A} = \frac{5.6 \cdot 10^5 \, \text{kg} \cdot 9.81 \frac{\text{m}}{\text{s}^2} - 1.244.000 \text{N} \cdot 0.1736}{\frac{1.225 \frac{\text{kg}}{\text{m}^3}}{2} \cdot \left(69.44 \frac{\text{m}}{\text{s}}\right)^2 \cdot 845 \text{m}^2} = 2.11.$$

The maximum lift coefficient $c_{A,\,max}$ is achieved by extending the slats and wing flaps.

For comparison, with the flat plate at an angle of attack of $\alpha = 20°$ (4.82b), a lift coefficient of

$$c_{A,plate} = 2\pi \sin\alpha = 2\pi \cdot 0.342 = 2.15$$

is reached.

5.13   Task: Water Jet Pump

Figure 5.18 shows the principle of a water jet pump. With the propelling jet of velocity c_0, fluid with velocity c_1 is sucked out of the ring cross-section. In the area of (1) ($\rightarrow$ 2) the two fluid streams mix and at the point (2) the liquid exits with the constant velocity c_2 as a free jet into the environment with the pressure p_a.

The wall friction can be neglected compared to the mixing losses of the two jets. The flow is assumed to be stationary and the medium incompressible. The pressure p_1 is constant over the entire cross-section in (1).

Given: $d_0 = 0.05\,\text{m}$, $d_1 = d_2 = 0.1\,\text{m}$, $\varrho = 1000\,\text{kg/m}^3$, $c_0 = 22\,\text{m/s}$, $c_1 = 3\,\text{m/s}$, $p_a = 10^5\,\text{Pa}$.

Wanted:

(a) What are velocity c_2 and mass flow $\dot{m}_2$ in exit cross section (2)?
(b) What is the pressure p_1 in the cross section (1)?
(c) What are the energy fluxes $\dot{E}_0, \dot{E}_1, \dot{E}_2$ and losses of $\dot{E}_V$ (1) $\rightarrow$ (2)?

Solution:

(a) Mass Conservation

$$\dot{m}_0 + \dot{m}_1 = \dot{m}_2 \rightarrow c_2 = \frac{d_0^2}{d_1^2}\cdot c_0 + \left(1 - \frac{d_0^2}{d_1^2}\right)\cdot c_1 = 7.750\,\text{m / s},$$

$$\dot{m}_2 = 60.868\,\text{kg / s}.$$

(b) Conservation of momentum

$$\varrho \cdot A_0 \cdot c_0^2 + \varrho \cdot (A_1 - A_0)\cdot c_0^2 - \varrho \cdot A_1 \cdot c_2^2 + p_1 \cdot A_1 - p_2 \cdot A_1 = 0$$

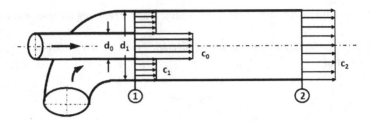

Fig. 5.18 Water jet pump

with $p_2 = p_a$ follows $p_1 = 32312.5\,\text{Pa}$.
(c) Energy conservation (power balance)

$$\dot{E}_0 + \dot{E}_1 = \dot{E}_2 + \dot{E}_V$$

$$\rightarrow \dot{m}_0 \cdot \left(\frac{p_1}{\varrho} + \frac{c_0^2}{2}\right) + \dot{m}_1 \cdot \left(\frac{p_1}{\varrho} + \frac{c_1^2}{2}\right) = \dot{m}_2 \cdot \left(\frac{p_2}{\varrho} + \frac{c_2^2}{2}\right) + \dot{m}_2 \cdot \frac{\Delta p_V}{\varrho}$$

$$\dot{E}_0 = 11849.449\,\text{W}, \dot{E}_1 = 650.539\,\text{W},$$

$$\dot{E}_2 = 7914.789\,\text{W}, \dot{E}_V = \dot{E}_0 + \dot{E}_1 - \dot{E}_2 = 4585.191\,\text{W}.$$

5.14   Task: Anemometer

The cup cross anemometer, as shown in Fig. 5.19, is a frequently used device for determining wind speed w_∞. Building services engineering is a major field of application for controlling façade elements. The advantage lies in the linear relationship between rotation speed and wind speed.

  Given: $r = 0.05\,\text{m}$, $R = 0.5\,\text{m}$, $c_{w1} = 1.33$, $c_{w2} = 0.34$, $\varrho = 1.225\,\text{kg/m}^3$.
  Wanted:

(a) The relationship between torque and the given geometric quantities r, R, the drag coefficients c_{w1}, c_{w2}, the density of the air, the incident flow velocity w_∞ and the rotational velocity $R\,\omega$ is to be represented dimensionless. A characteristic velocity parameter λ follows from the ratio of the rotational velocity $R \cdot \omega$ to the incident flow velocity w_∞.
(b) The torque M_D is to be determined via a moment balance and plotted in dimensionless form M_D/M_N with $M_N = \varrho / 2 \cdot w_\infty^2 \cdot R^2$, over the velocity parameter λ.

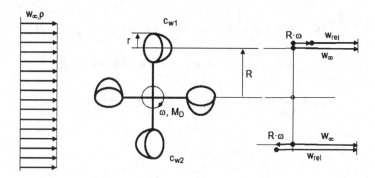

Fig. 5.19 Principle sketch of the wind anemometer

(c) The power P is to be normalized with $P_N = \varrho / 2 \cdot w_\infty^3 \cdot R^2$ and represented as a function of λ.

(d) Discuss the three cases:
1. $\omega = 0$ locked rotor,
2. $M_D = 0$ anemometer function and
3. $P = P_{max}$ the maximum power when operating as a wind turbine.

Solution:

(a) $M_D = f(\text{geometry}, \text{media}, \text{BC}) = f(r, R, c_{w1}, c_{w2}, \varrho, w_\infty, \omega)$

With the dimensional analysis, 5 dimensionless parameters follow from these 8 influencing variables minus the 3 basic variables (M, L, T):

$$c_M = \frac{M_D}{\dfrac{\varrho}{2} w_\infty^2 R^3}, \ \frac{r}{R}, \ \lambda = \frac{R \cdot \omega}{w_\infty}, \ c_{w1}, \ c_{w2} \ \rightarrow \ c_M = F\left(\frac{r}{R}, \lambda, c_{w1}, c_{w2}\right).$$

(b) Moment balance

$$\frac{\varrho}{2}\left(w_\infty - R\cdot\omega\right)^2 \pi\cdot r^2 \cdot c_{w1}\cdot R = \frac{\varrho}{2}\left(w_\infty + R\cdot\omega\right)^2 \pi\cdot r^2 \cdot c_{w2}\cdot R + M_D$$

$$M_D = \frac{\varrho}{2}\pi\cdot r^2 \cdot R\cdot\left[\left(w_\infty - R\cdot\omega\right)^2 \cdot c_{w1} - \left(w_\infty + R\cdot\omega\right)^2 \cdot c_{w2}\right]$$

$$c_M = \pi\cdot\left(\frac{r}{R}\right)^2 \cdot\left[\left(1-\lambda\right)^2 \cdot c_{w1} - \left(1+\lambda\right)^2 \cdot c_{w2}\right]$$

$$\frac{P}{P_N} = \frac{\dfrac{\varrho}{2}w_\infty^2 R^3 \omega\cdot c_M}{\dfrac{\varrho}{2}w_\infty^3 R^3} = c_M\cdot\frac{R\omega}{w_\infty} = c_M\cdot\lambda$$

(c) **Case (1):** When the rotor is held stationary ($\lambda = 0$), the moment is maximum. It then decreases linearly with increasing parameter λ, as shown in Fig. 5.20. **Case (2):** If the rotor is not loaded, the resulting moment is $M_D = 0$. It is then

$$\left(1-\lambda\right)^2 \cdot c_{w1} - \left(1+\lambda\right)^2 \cdot c_{w2} = 0 \rightarrow \sqrt{\frac{c_{w2}}{c_{w1}}} = \frac{1-\lambda}{1+\lambda} \rightarrow \lambda = 0.3284$$

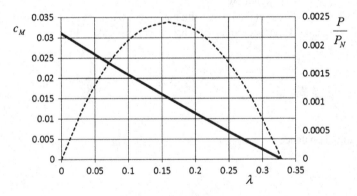

Fig. 5.20 Course of torque (*solid*) and power (*dashed*) as a function of λ

Fig. 5.21 Characteristic
speed n as a function of wind
speed w_∞

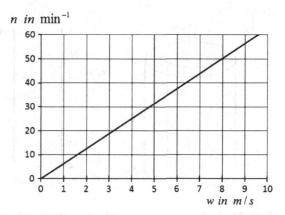

From $\lambda = \dfrac{R \cdot \omega}{w_\infty}$ follows $\omega = 2\pi \cdot n = \dfrac{w_\infty \cdot \lambda}{R}$, which corresponds to a linear relationship (characteristic curve Fig. 5.21) between the wind speed w_∞ and the
angular velocity ω or the rotational speed n.

Case (3): The power P has the maximum value for the value $\lambda = 0.159$ as can be
seen from Fig. 5.20. With the given data, a wind speed of $w_\infty = 10$ m/s results in
a maximum power of $P = 0.370$ W. Wind turbines of this type are therefore only
conditionally suitable for energy conversion.

5.15   Task: Vortex Flow in the Cylinder Gap

The flow in the gap between two rotating cylinders is to be analyzed with respect
to the transport properties of momentum and energy in the radial direction and the
associated dissipation (frictional heat). The flow is rotationally symmetric and is
said to have only the circumferential velocity $v(r)$, while the radial and axial velocities are zero. The pressure distribution $p(r)$ is obtained by the relation (4.24).
Figure 5.22 shows the gap geometry filled with a viscous incompressible fluid
($\eta = \varrho \cdot \nu$). The velocity distribution $v(r) = A \cdot r + B/r$ is composed of a rigid-body
vortex and a potential vortex. The constants A and B are given with the boundary
conditions $v(R_1) = \omega_1 \cdot R_1$ and $v(R_2) = \omega_2 \cdot R_2$.

Given: R_1, R_2, ℓ, ϱ, ν, ω_1, ω_2.

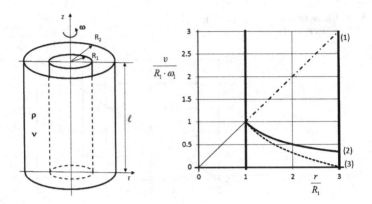

Fig. 5.22 Cylinder gap geometry and velocity distributions in the cylinder gap

Wanted:

(a) The velocity distribution $v(r)$ with constants A and B.

(b) The shear stress distribution $\tau(r) = -\eta \cdot \left[r \cdot \dfrac{\partial}{\partial r}\left(\dfrac{v}{r}\right) \right] = -\eta \cdot \left(\dfrac{dv}{dr} - \dfrac{v}{r}\right)$.

(c) The torques M_1 and M_2 on the inner and outer cylinder.

(d) The dissipation $\Phi = L_1 - L_2$ is to be determined from the power balance with

$$L = \frac{dE}{dt} = \dot{E} \quad \text{as well as from the volume integral} \quad \Phi = \iiint\limits_{v} \eta \cdot \left[r \cdot \frac{\partial}{\partial r}\left(\frac{v}{r}\right) \right]^2 \cdot dV.$$

(e) The limit cases

    (1) $v(r) = B/r$ with $A = 0$,

    (2) $v(r) = A \cdot r$ with $B = 0$,

    (3) $v(r) = A \cdot r + B/r$ with $\omega_2 = 0$,

    (4) $R_2 \to \infty$, $v(r \to \infty) = 0$,

are to be analyzed.

Solution:

(a) From $v(r) = A \cdot r + B/r$ with $v(R_1) = \omega_1 \cdot R_1$ and $v(R_2) = \omega_2 \cdot R_2$ follow the constants

$$A = \frac{\omega_2 \cdot R_2^2 - \omega_1 \cdot R_1^2}{R_2^2 - R_1^2}, \quad B = \frac{R_1^2 \cdot R_2^2 \cdot (\omega_1 - \omega_2)}{R_2^2 - R_1^2}.$$

(b) The shear stress distribution thus follows

$$\tau(r) = -\eta \cdot \left(\frac{dv}{dr} - \frac{v}{r} \right) = \eta \cdot \frac{2 \cdot B}{R_1^2}, \quad B = \frac{R_1^2 \cdot R_2^2 \cdot (\omega_1 - \omega_2)}{R_2^2 - R_1^2}$$

(c) The torque at the inner cylinder follows from the integration over the cylinder surface area to

$$M_1 = R_1 \cdot \int_0^\ell \int_0^{2\pi} \tau(R_1) \cdot R_1 \cdot d\varphi \cdot dz$$

$$= 4 \cdot \pi \cdot \eta \cdot \ell \cdot B = 4 \cdot \pi \cdot \eta \cdot \ell \cdot \frac{R_1^2 \cdot R_2^2 \cdot (\omega_1 - \omega_2)}{R_2^2 - R_1^2},$$

and from the moment equilibrium then follows for M_2 at the outer cylinder the same value, but with opposite direction.

(d) For the dissipation Φ, the following results from the power balance and from the volume integral over the local dissipation

$$\Phi = L_1 - L_2 = M_1 \cdot \omega_1 - M_2 \cdot \omega_2$$

$$= \int_0^\ell \int_{R_1}^{R_2} \int_0^{2\pi} \eta \cdot \frac{4 \cdot B^2}{r^4} \cdot r \cdot d\varphi \cdot dr \cdot dz = 4 \cdot \pi \cdot \eta \cdot \ell \cdot \frac{R_1^2 \cdot R_2^2 \cdot (\omega_1 - \omega_2)^2}{R_2^2 - R_1^2}.$$

The difference between the power L_1 supplied at the inner cylinder and the power L_2 discharged at the outer cylinder corresponds to the dissipation Φ.

(e) **Limit case (1):** For the condition, $\omega_2 \cdot R_2^2 - \omega_1 \cdot R_1^2 = 0$ the constant $A = 0$ and the potential vortex is established in the cylinder gap. The viscous gap flow is then rotation-free and for the dissipation follows

$$\Phi = 4 \cdot \pi \cdot \eta \cdot \ell \cdot \frac{R_1^2 \cdot R_2^2 \cdot (\omega_1 - \omega_2)^2}{R_2^2 - R_1^2}.$$

A viscous potential flow is then present in the cylinder gap.
Limit case (2): For the case that the angular velocity is $\omega_1 = \omega_2$, a rigid body rotation is established in the gap. The medium rotates like a rigid body and does not flow. No shear stresses occur.

Limit case (3): If the outer cylinder is held stationary, the general form of the velocity distribution is a combination of rigid-body and potential vortex. In this case, an unstable centrifugal force layering is present. As soon as the critical Taylor number Ta is exceeded, toroidal vortices set in. For small gap widths $s = R_2 - R_1$ and $s/R_1 \ll 1$, this Taylor number has the value $Ta^2 = \dfrac{R_1 \omega_1 s^3}{\nu^2} = 1708$.

Limit case (4): In this case, the situation is that of a rotating cylinder immersed in a viscous medium. The circumferential speed corresponds to that of the potential vortex. The power supplied to the rotating cylinder (energy flow) corresponds to the dissipation (frictional heat) generated in the medium. In this case, as in (1), there is also a viscous potential flow.

5.16   Task: Source/Sink Flow in the Cylinder Gap

The steady incompressible flow between two porous cylinders in Fig. 5.23 is to be analysed. Only the radial velocity u occurs, while the circumferential and axial velocities are zero.

The mass and momentum balance for radial flow in cylindrical coordinates is:

Fig. 5.23 Radial source flow in the cylinder gap

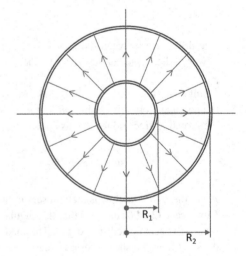

$$0 = \frac{1}{r} \cdot \frac{\partial}{\partial r}(r \cdot u)$$

$$u \cdot \frac{\partial u}{\partial r} = -\frac{1}{\varrho} \cdot \frac{\partial p}{\partial r} + v \cdot \left[\frac{\partial}{\partial r} \left(\frac{1}{r} \cdot \frac{\partial}{\partial r}(r \cdot u) \right) \right]$$

Gravity in the z-direction is eliminated, which is possible for incompressible media. The pressure p thus corresponds to the superposition to the hydrostatic pressure distribution.

Given: R_1, R_2, ℓ, ϱ, ν, $\dot{V}$.

Wanted:

(a) The velocity distribution $u(r)$.
(b) The pressure distribution $p(r)$.
(c) Discuss the solutions of (a) and (b) for $u(r) < 0$ sink flow.
(d) The dissipation $\Phi = L_1 - L_2$ can be calculated from the power balance as well as via the local dissipation ϕ with the volume integral

$$\Phi = \iiint_V \phi \cdot dV = \iiint_V 2 \cdot \eta \cdot \left[\left(\frac{\partial u}{\partial r} \right)^2 + \left(\frac{u}{r} \right)^2 \right] \cdot dV.$$

Solution:

(a) From the mass balance follows for the radial velocity

$$r \cdot u(r) = C \quad \rightarrow \quad u(r) = \frac{C}{r} \quad \text{with } C = \frac{\dot{V}}{2 \cdot \pi \cdot \ell}$$

where the constant C follows from the constancy of the volume flow $\dot{V} = 2 \cdot \pi \cdot r \cdot \ell \cdot u(r)$.

(b) With the result of continuity, the friction term in the momentum balance is omitted.

The pressure distribution $p(r)$ follows by integration to

$$\frac{dp}{dr} = -\varrho \cdot u \cdot \frac{du}{dr} = \varrho \cdot \frac{C^2}{r^3} \quad \rightarrow \quad p(r) = p(R_1) + \frac{\varrho \cdot C^2}{2} \left(\frac{1}{R_1^2} - \frac{1}{r^2} \right)$$

(c) The velocity distribution $u(r)$ and the pressure distribution $p(r)$ are independent of direction. They apply equally to the radial source and sink flow.
This exact solution of the Navier-Stokes equation is independent of the viscosity ν. However, the dissipation Φ is different from zero.
This flow is rotation-free, it has a potential and therefore called viscous potential flow.

(d) Dissipation is transmitted at the cylindrical boundary surfaces by the power of viscous normal stresses $\sigma_{rr} = -2 \cdot \eta \cdot \dfrac{\partial u}{\partial r}$.

$$\Phi = L_1 - L_2 = \iint_A u(r) \cdot \sigma_{rr} \cdot dA \bigg|_1 - \iint_A u(r) \cdot \sigma_{rr} \cdot dA \bigg|_2 = \frac{\eta \cdot \dot{V}^2}{\pi \cdot \ell} \left(\frac{1}{R_1^2} - \frac{1}{R_2^2} \right)$$

The evaluation via local dissipation provides the same value:

$$\Phi = \iiint_V \phi \cdot dV = \iiint_V 2 \cdot \eta \cdot \left[\left(\frac{\partial u}{\partial r} \right)^2 + \left(\frac{u}{r} \right)^2 \right] \cdot dV = \frac{\eta \cdot \dot{V}^2}{\pi \cdot \ell} \left(\frac{1}{R_1^2} - \frac{1}{R_2^2} \right)$$

Dimensions and Units of the Most Important Occurring Quantities

Size, designation	Dimensions		Units
	F, L, T, ϑ	M, L, T, ϑ	
Length	L	L	Meter, m
Force	F	MLT^{-2}	Newton, N
Mass	$FL^{-1}T^{2}$	M	Kilogram, kg
Time	T	T	Second, s
Temperature	ϑ	ϑ	Kelvin, K
Speed	LT^{-1}	LT^{-1}	m/s
Acceleration	LT^{-2}	LT^{-2}	m/s^2
Pressure, tension	FL^{-2}	$ML^{-1}T^{-2}$	Pascal, Pa = N/m^2
Moment, work, energy	FL	$ML^{2}T^{-2}$	Joule, J = Ws = Nm
Power, energy flow	FLT^{-1}	$ML^{2}T^{-3}$	Watt, W = Nm/s
Density ϱ	$FL^{-4}T^{2}$	ML^{-3}	kg/m^3
Mass flow $\dot{m}$	$FL^{-1}T$	MT^{-1}	kg / s
Dynamic viscosity η	$FL^{-2}T$	$ML^{-1}T^{-1}$	Pas = Ns/m^2
Kinetic viscosity ν	$L^{2}T^{-1}$	$L^{2}T^{-1}$	m^2/s
Coefficient of expansion β	ϑ^{-1}	ϑ^{-1}	1 / K
Specific heat c_p, c_v	$L^{2}T^{-2}\vartheta^{-1}$	$L^{2}T^{-2}\vartheta^{-1}$	J/kgK
Thermal conductivity λ	$FT^{-1}\vartheta^{-1}$	$MLT^{-3}\vartheta^{-1}$	W/mK
Surface tension σ	FL^{-1}	MT^{-2}	N/m
Thermal diffusivity $k = \lambda/\varrho\, c_p$	$L^{2}T^{-1}$	$L^{2}T^{-1}$	m^2/s
Heat transfer coefficient α	$FL^{-1}T^{-1}\vartheta^{-1}$	$MT^{-3}\vartheta^{-1}$	W/m^{2}K
Special gas constant R_i	$L^{2}T^{-2}\vartheta^{-1}$	$L^{2}T^{-2}\vartheta^{-1}$	J/kgK
Entropy s	$L^{2}T^{-2}\vartheta^{-1}$	$L^{2}T^{-2}\vartheta^{-1}$	J/kgK

© Springer Fachmedien Wiesbaden GmbH, part of Springer Nature 2022
J. Zierep, K. Bühler, *Principles of Fluid Mechanics*,
https://doi.org/10.1007/978-3-658-34812-0

Size, designation	Dimensions		Units
	F, L, T, ϑ	M, L, T, ϑ	
Dissipation, volume-related ϕ	$FL^{-2}T^{-1}$	$ML^{-1}T^{-3}$	W/m^3
Total dissipation Φ	FLT^{1}	$ML^{2}T^{-3}$	W

Literature

General Fluid Mechanics

1. Albring, W.: Angewandte Strömungslehre. 6. Aufl. Akademie Verlag, Berlin (1990)
2. Becker, E.: Technische Strömungslehre. 7. Aufl. Teubner, Stuttgart (1993)
3. Böswirth, L., Bschorer, S.: Technische Strömungslehre. 10. Aufl. Springer Vieweg, Wiesbaden (2014)
4. Eck, B.: Technische Strömungslehre. 8. Aufl. Springer, Berlin/Heidelberg/New York (1978)
5. Gersten, K.: Einführung in die Strömungsmechanik. Shaker, Aachen (2003)
6. Herwig, H.: Strömungsmechanik. 2. Aufl. Springer Vieweg, Wiesbaden (2016)
7. Leiter, E.: Strömungsmechanik nach Vorlesungen von K. Oswatitsch. Vieweg. Braunschweig (1978)
8. Oertel, H. Jr., Böhle, M., Reviol, T.: Strömungsmechanik. 7. Aufl. Springer Vieweg, Wiesbaden (2015)
9. Oertel, H. Jr., Böhle, M., Reviol, T.: Übungsbuch Strömungsmechanik. 8. Aufl. Vieweg+Teubner, Wiesbaden (2012)
10. Oswatitsch, K.: Physikalische Grundlagen der Strömungslehre. Handbuch der Physik, Bd. VIII/1. Springer, Berlin/Heidelberg/New York (1959)
11. Oertel, H.: Prandtl – Führer durch die Strömungslehre. 14. Aufl. Springer Vieweg, Wiesbaden (2017)
12. Schade, H., Kameier, F., Kunz, E., Paschereit, C. O.: Strömungslehre. 4. Aufl. de Gruyter, Berlin/New York (2013)
13. Truckenbrodt, E.: Fluidmechanik I, II. 4. Aufl. Springer, Berlin (2008)
14. Wieghart, K.: Theoretische Strömungslehre. 2. Aufl. Universitätsverlag, Göttingen (2005)

© Springer Fachmedien Wiesbaden GmbH, part of Springer Nature 2022
J. Zierep, K. Bühler, *Principles of Fluid Mechanics*,
https://doi.org/10.1007/978-3-658-34812-0

15. Zierep, J., Bühler, K.: Strömungsmechanik. In: Hennecke M., Strotzki B.(eds) Hütte –
    Das Ingenieurwissen. 35. Aufl. Springer, Berlin/Heidelberg/New York (2019)
16. Zierep, J., Bühler, K.: Strömungsmechanik. Springer, Berlin/Heidelberg/New York
    (1991)

Subfields of Fluid Mechanics

17. Becker, E.: Gasdynamik. Teubner, Stuttgart (1969)
18. Keune, F., Burg, K.: Singularitätenverfahren der Strömungslehre. Braun, Karlsruhe
    (1975)
19. Laurien, E., Oertel, H. Jr.: Numerische Strömungsmechanik. 5. Aufl. Springer Vieweg,
    Wiesbaden (2013)
20. Oswatitsch, K.: Grundlagen der Gasdynamik. Springer, Wien (1976)
21. Schlichting, H., Gersten, K.: Grenzschichttheorie. 10. Aufl. Springer, Berlin (2006)
22. Schlichting, H., Truckenbrodt, E.: Aerodynamik des Flugzeuges. 2 Bde. 3. Aufl. Springer,
    Berlin (2001)
23. Schneider, W.: Mathematische Methoden der Strömungsmechanik. Vieweg, Braunsch-
    weig (1978)
24. Zierep, J.: Ähnlichkeitsgesetze und Modellregeln der Strömungslehre. 3. Aufl. Braun,
    Karlsruhe (1991)
25. Zierep, J.: Strömungen mit Energiezufuhr. 2. Aufl. Braun, Karlsruhe (1990)
26. Zierep, J.: Theoretische Gasdynamik. 4. Aufl. Braun, Karlsruhe (1991)

Index

© Springer Fachmedien Wiesbaden GmbH, part of Springer Nature 2022
J. Zierep, K. Bühler, *Principles of Fluid Mechanics*,
https://doi.org/10.1007/978-3-658-34812-0

Printed in the United States
by Baker & Taylor Publisher Services

Printed in the United States
by Baker & Taylor Publisher Services